Solar Radiation

Solar Radiation

Edited by **Catherine Waltz**

New York

Published by Callisto Reference,
106 Park Avenue, Suite 200,
New York, NY 10016, USA
www.callistoreference.com

Solar Radiation
Edited by Catherine Waltz

International Standard Book Number: 978-1-63239-571-9 (Hardback)

Printed in the United States of America.

Contents

Preface

This book is the result of concerted efforts by the eminent scientists and experts in the field of solar radiation from all over the world. It discusses ecological impacts of solar radiation. The book covers numerous topics on solar radiation such as architectural application, thermal energy application and electricity generation application. It provides scientific understanding on solar radiation for reference to researchers and students.

This book is a result of research of several months to collate the most relevant data in the field. When I was approached with the idea of this book and the proposal to edit it, I was overwhelmed. It gave me an opportunity to reach out to all those who share a common interest with me in this field. I had 3 main parameters for editing this text:

1. Accuracy – The data and information provided in this book should be up-to-date and valuable to the readers.

2. Structure – The data must be presented in a structured format for easy understanding and better grasping of the readers.

3. Universal Approach – This book not only targets students but also experts and innovators in the field, thus my aim was to present topics which are of use to all.

Thus, it took me a couple of months to finish the editing of this book.

I would like to make a special mention of my publisher who considered me worthy of this opportunity and also supported me throughout the editing process. I would also like to thank the editing team at the back-end who extended their help whenever required.

Editor

Section 1

Architectural Application

Solar Radiation in Buildings, Transfer and Simulation Procedures

Jose Maria Cabeza Lainez
University of Seville
Spain

1. Introduction

Solar radiation is the only renewable energy source readily available at every building in the world. Whilst urban regulations and meteorological or geographical factors often impede proper ventilation, to design a building without at least a view of the surroundings is tantamount to making plans for a prison or a tomb, and no culture would accept that as a permanent residence. Thus, the necessary connections with the environment are provided by apertures through which radiation is admitted.

Since antiquity, a multitude of researchers and scientists have striven to find the magnitude of solar radiation incident on a horizontal surface at the earth's crust. A few of them have found adroit correlations between horizontal and vertical irradiation. These seem acceptable for the analysis of building facades since direct measurements are in many cases not feasible due to obstructions, interferences with ground reflection or simply because of economic constraints.

However, it is still surprising how few scholars are familiar with the distribution of such radiation inside the chambers, precisely where it should be used. To be sure, if one wants to transfer a certain amount of energy to human beings, the task needs to be accomplished piecemeal, or the consequences could be devastating as, unfortunately, everybody knows.

In the ensuing chapter, the author intends to explain the fundamentals of radiative energy transfer, a discontinued branch of geometric optics that colligates time, space and architecture in a single operation. The author would also try to ensure that every person is able to reproduce his experiments at home by virtue of computer simulation and analysis.

2. Radiative transfer between spherical surfaces

Let us start by discussing radiative exchanges in simple volumes. The sphere is a volume enclosed by only one surface and with some restrictions it is used both in engineering and architecture. If the inner surface of the sphere emits under Lambertian diffusion, the total fraction of energy reaching the same surface will be one hundred percent, that is: the unity.

In mathematical terms this is expressed as:

$$F_{11} = 1 \qquad (1)$$

The fraction of energy leaving surface 1 and arriving at surface 1, is one. The former gives rise to a new algebra defined by:

$$F_{i1} + F_{i2} + ... + F_{in} = 1 \tag{2}$$

In a closed volume, by the principle of energy conservation, and disregarding transmission losses, radiant energy emitted by surface i is necessarily distributed in its entirety among the surrounding surfaces. As an example, in a cube, where all faces are equal, the fraction of energy leaving from an internal source to any of the other five is $1/5=0.2$.

As the radiant flux is originated in a given surface and bears only nominal relationship with the medium in which the phenomena take place, the distribution of such flux will be a function of the dimensions of the surfaces involved. Thus we could anticipate a second and final property for our algebra (Lambert, 1764).

$$A_i * F_{ij} = A_j * F_{ji} \tag{3}$$

Where A_i is the area of surface i and F_{ij} is the fraction of energy that leaves i and reaches j.

Returning to the sphere, it is easy to see that although interior radiation may occur, unless we are able to pierce the surface by means of some kind of section, interaction with the environment remains negligible. Any planar section of the sphere will produce a spherical cap with a circular base that works as an aperture. (See Figure 1).

The size of the aperture is determined by the height of the cap, h. Beginning with the case of a hemisphere, the said height coincides with the radius of the sphere $R=a=h$.

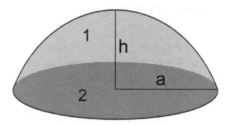

Fig. 1. Surfaces generated by a spherical cap

In this situation, there are two surfaces involved in the radiation problem. Let us apply the former algebra to them.

If for the whole sphere $F_{11} = 1$, taking into account the dependence on area of radiative transfer, it is assumed that for the hemisphere $F_{11} = 1/2 = 0.5$, and the demonstration will be given later in the text.

By equation (2) the former also implies that $F_{12} = 0.5$. Thus, half of the flux of the hemisphere is distributed over itself and the other half is ceded to the circular base that serves as an aperture of the sphere.

In accordance with this reasoning, for a cap whose area equals one third (1/3) of the whole surface, $F_{11} = 1/3$ and $F_{12} = 2/3$ and so on.

When the respective areas of surfaces 1 and 2 are introduced, the fraction of energy that leaves 1 and arrives at 2 equates the area-ratio of such potential sources. In this way, equation 3 is proved if we remember that, for the spherical cap, F_{21} has to be one. The circle is a planar figure and gives one hundred percent of its energy to the surrounding cap.

$$F_{12} = \frac{a^2}{a^2 + h^2} \tag{4}$$

It is inferred that $F_{11} = 1 - F_{12}$, and its value would be,

$$F_{11} = \frac{h^2}{a^2 + h^2} \tag{5}$$

Substituting into equation 5, the trigonometric relation for the radius of the sphere, R,

$$a^2 + h^2 = 2 * R * h \tag{6}$$

We obtain an extremely important and beautiful expression,

$$F_{11} = \frac{h}{2 * R} \tag{7}$$

Thus, by simple logic, and with hardly any calculus, the author has solved for the first time in this field one of the most complex integral equations of environmental science (Eq. 8).

$$\emptyset_{1-2} = E_{b1} \int_{A_1} \int_{A_2} cos\theta_1 * cos\theta_2 * \frac{dA_1 * dA_2}{\pi * r^2} = E_{b1} * \pi * a^2 \tag{8}$$

Where Φ_{1-2} is the radiative flux exchange between the surfaces considered, and E_{b1} is the radiant energy emitted by surface 1. The relative ease of the solution is partly due to the fact that the quantities termed $cos\theta_i$, which represent the cosines of the angles between the line going from the center of surface 1 to the center of surface 2 and their respective normals, are simpler to find in this case as the said normals always pass through the center of the sphere.

It is convenient to use the former results in an ample variety of ways.

For instance, if the aforementioned division of the sphere is performed by means of two planar sections, like in a quarter of sphere (see Figure 2), the solutions are still valid, in the sense that the quarter of a sphere is giving itself one fourth of its emissive power $F_{33} = 1/4$.

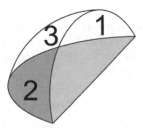

Fig. 2. Surfaces generated by a quarter of a sphere

As we already found, the two semicircles receive in total 3/4 of the flux, but provided that they are of equal area, each one of them receives $3/8 = F_{32} = F_{31}$.

Thus, equation 2 is fulfilled.
By equation 3, the so-called principle of reciprocity, $F_{13} = F_{23} = 2*\pi*R^2/\ \pi*R^{2*}\ F_{31} = 3/4$
And this implies that $F_{12} = F_{21} = 1/4$
A second complex integral equation has been solved by the author without calculus.

In a similar manner, adjusting the fragment of sphere which the problem may demand, the radiative exchange between semicircles with a common edge, forming any angle from 0 to 180 degrees, can be found. The above example is valid for 90 degrees.

The author has been the first to propose the following equation, previously unheard of in the literature, to obtain the energy balance between the said semicircles, where x represents the value of the internal angle (Fig. 3),

$$F_{12} = 1 - \frac{x}{90} + \frac{x^2}{32400} \qquad (9)$$

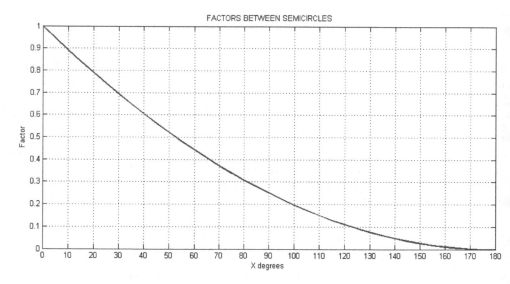

Fig. 3. Radiative exchanges between two semicircles with a common edge and forming an internal angle x

So far so good. The former expression solves a whole set of integral equations and anybody can understand how the radiant flux is transferred from circular or semicircular apertures to the interior of the sphere as a total fraction. However, it is often useful to examine this transfer in more detail, i.e. point by point.

2.1 Radiative transfer between spherical surfaces and points

Referring again to the sphere in respect with the canonical equation 8; if you look at figure 4, it is easy to find the relationship between r, $cos\theta$ and the radius of the sphere R.

$$\emptyset_{1-2} = \frac{E_{b1}}{4*\pi*R^2} \int_{A_1} \int_{A_2} dA_1 * dA_2 \tag{10}$$

$4*\pi*R^2$ is obviously the area of the sphere. Thus, the radiative flux transfer is dependent on the size of the surfaces but not on their position on the sphere and for a given area it is also a constant.

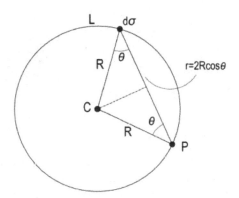

Fig. 4. Differential surfaces in the sphere used to find the radiative exchage

Those properties are unique to the spherical surface and also crucial for our discussion.

At this point, if we return to the spherical cap described in figure 1, there the flux from the cap to the circle was defined. However, assuming this circle to be virtual or, in other words, transparent to radiation, what remains behind is simply the opposite cap of the sphere, now called surface 2 for convenience.

The fraction of energy emitted from 1 to the new surface 2 is the same as in equation 4, but the energy received by surface 2 is F_{21}, and we can obtain it from the theorem of reciprocity finding the ratio between the respective areas.

The area of surface 2 is,

$$A_2 = \pi * (a^2 + (2 * R - h)^2) \tag{11}$$

$$F_{21} = \frac{\pi*a^2}{\pi*(a^2+(2*R-h)^2)} \tag{12}$$

And from equation 6,

$$a^2 + h^2 = 2 * R * h \tag{13}$$

$$F_{21} = \frac{a^2*h^2}{a^4+2*h^4-2*h^4+3*a^2*h^2-2*a^2*h^2} \tag{14}$$

$$F_{21} = \frac{a^2*h^2}{a^2(a^2+h^2)} = \frac{h^2}{a^2+h^2} = \frac{h}{2*R} \tag{15}$$

This equates the fraction defined as F_{11} for the spherical cap. Let us see it depicted in a different graph (Figure 5).

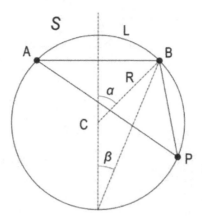

Fig. 5. Cap and sphere where the radiative exchange takes place

In figure 5, the cap S goes from A to B and its area is $\pi*2*R*h$, as mentioned above (Eq. 6). Dividing the former by the total area of the sphere according to equation 10, the result is as shown in Equations 7 and 15. That is, the energy received at any point of the interior sphere wall outside the spherical cap S is constant, $h/2*R$. This is often expressed as $1/2*(1-\cos\alpha)$ or $1/2*(1-\cos2\beta)$.

With the former property we can replace the cap source by its enclosed circle AB.

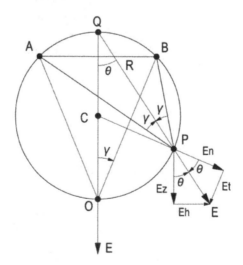

Fig. 6. The radiation vector in a sphere

Radiant energy due to the spherical cap or the said circle, in a direction normal to the interior of the sphere, is constant but it is mandatory not to forget that radiation is in truth a vector, meaning that its projection on different planes may present diverse values.

Finding the radiation vector that originates in the cap source poses no particular problem (See Figure 6). For reasons of symmetry, its centre has to be at point Q, following the well-known principle of the circumference by virtue of which, equal arcs subtend equal angles[1].

The extreme of the vector lies at the point under study. If the direction of the vector is known, it only remains to calculate its modulus. Using trigonometric properties, as the normal E_N is constant, this implies that the vertical component E_Z equates the normal for the value of direction angle θ is a half of the angle subtended by the arc OP from the centre of the sphere, being P the point under study and O the horizontal projection of the centre.

The last step to determine the modulus of the vector is to project its vertical component $h/(2*R)$ onto the horizontal plan, multiplying by the tangent of θ.

The fact that the vertical component is constant has led to the construction of useful graphs in which horizontal radiation E_Z at a given point is obtained as a function of the radius of the cap's base and the distance from the circle's centre to the point considered (See figures 7 and 8).

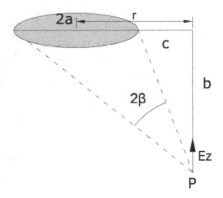

Fig. 7. Perpendicular component of the radiation vector under a disk

If the said quantities are known, the coordinate component E_H, which is constant only for the same height over the horizontal in the sphere, in other words, by parallels, can be found employing the circumference's properties.

$$R^2 = a^2 + \left(\frac{b^2+r^2-a^2}{2b}\right)^2 \tag{16}$$

$$h = R - \sqrt{R^2 - a^2} \tag{17}$$

And the vertical distance from the origin of the radiation vector to the point considered is, evidently,

$$d = b + h \tag{18}$$

---

[1] Both MacAllister and Sumpner failed to see this point and located the origin of the vector at the centre of the enclosed circle, though this error may be relevant for sizeable sources.

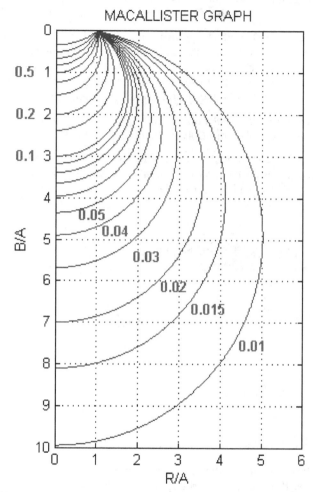

Fig. 8. The MacAllister graph for finding the vertical component of radiation, using only the magnitudes on figure 7 and without need for calculation

To obtain the vectorial field of radiation around a spherical cap source or its equivalent disk (the final aim of radiative transfer and simulation for such common geometry) is tantamount to finding a set of virtual spheres which contains both the source and each one of the points considered in the reticule that represents the field. From this set we extract the components and consequently the radiation vector at each point of the domain analysed.

The author has created original software that simulates the aforementioned fields for planes, spheres and cylinders at every possible inclination angle in spherical coordinates, θ or φ; some results are presented in figures 9 and 10.

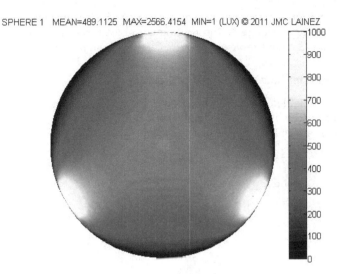

Fig. 9. Distribution of radiation on a sphere of radius 20 m. and reflectance 0.3, under three circular sources of 9 m. diameter and 10000 lumen/m^2 (lux) intensity, rotated 120°

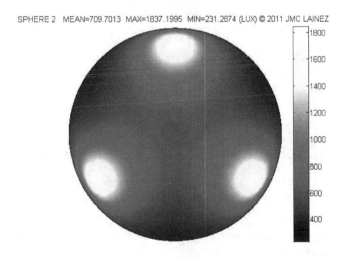

Fig. 10. The same three sources as in fig. 9 but radiation values are found instead on a circular disk 7.5 metres under the centre of the sphere

On the understanding that radiation fields are additive due to their vectorial nature, the author has solved with ease the fundamental problem of radiative transfer. To be sure, not all the openings[2] in buildings are circular but a significant amount of them can be approximated to one or several emitting disks with sufficient precision, taking into account

---

[2] The word "open" comes from ancient Greek οπή (opeh), meaning eye. The main aperture or vent in classic buildings is often termed "opaion" or "oculus", its latin equivalent.

the inevitable lack of accuracy in the construction industry. In any case, the author has adapted his software to triangular and rectangular apertures (see section 4) but in the latter situation, solutions are not simple and require complex integration[3].

2.2 Radiative transfer between complex surfaces

Once this matter is settled, retrieving the fundamental equation expounded at 8 to be partly solved in 10 and extending it to a second spherical cap as in figure 11, the radiant flux between the two caps of heights h_1 and h_2 is,

$$\emptyset_{1-2} = E_{b1} * \pi * h_1 * h_2 \qquad (19)$$

For the area of the second cap is nothing but $\pi*2*R*h_2$.

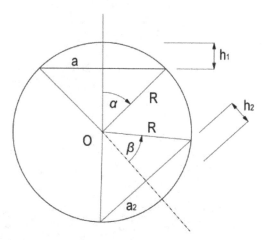

Fig. 11. Two spherical caps inside a sphere employed to find the radiative transfer

In the special situation that $h_1 = h_2 = h$, which often coincides with parallel disks of equal radius a, the flux would be $E_{b1}*\pi*h^2$ and the fraction of energy from disk 1 to disk 2 (or their surrounding caps), equates h^2/a^2.

If only the perpendicular distance between the disks, called $2b$, is known (see figure 12), the height of the cap would be,

$$h = \sqrt{a^2 + b^2} - b \qquad (20)$$

Thus, the fraction is obtained as,

$$F_{12} = F_{21} = \frac{a^2 + 2*b^2 - 2*b*\sqrt{a^2+b^2}}{a^2} \qquad (21)$$

---

[3] As a matter of fact only J. H. Lambert was capable of finding a solution for perpendicular rectangles with a common edge without the help of integration, but he often complained in his book of the "insurmountable difficulties" that the process entailed.

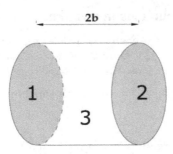

Fig. 12. Surfaces defined by a cylindrical volume used to find the radiative transfer

With the former expression, solving the radiative transfer inside cylinders is easy as there are only three surfaces involved and we can form a non-trivial system of two equations with two unknowns. Changing the circular base of the cylinder by a spherical cap (fig.13) will alter some values but not the general problem because all the possibilities of transfer between caps and disks have already been explored. For instance, if the cap is a hemisphere, the values of the factors to the disk and the cylinder need to be affected by $a^2/(a^2 +h^2)=0.5$ (Eq. 4), and progressively until reaching one which is again the planar disk. [4]

The resulting volume in figure 13 has been used by humankind for centuries; mainly in churches but also in libraries, concert halls, banks, markets or pavilions of any sort. With another cap on the base, the form is more recently used for silos, fluid reservoirs and containment vessels at power stations.

Having solved the primary transference problems, the following step is the important subject of interreflections.

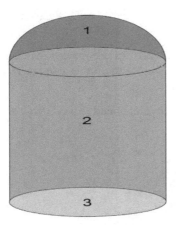

Fig. 13. Volume composed of a cylinder and a spherical cap used to find the radiative transfer among those surfaces

---

[4] Note that values under 0.5 can also be found for this relationship in a sort of globular cap with an area bigger than the hemisphere.

3. Interreflections among surfaces in a volume

Thus far, we have considered that energy is transferred from a primary source to several secondary ones. However this procedure does not deal with the fact that the receiving sources, being partially absorptive, may in turn become emitters.

In this situation, the total balance of energy is obtained by equation 22,

$$E_{tot} = E_{dir} + E_{ref} \tag{22}$$

Where E_{dir} represents the energy received directly and E_{ref} is the reflected energy. These two terms added give the total balance of radiative energy E_{tot}. When more than one surface is involved, expression 22 generates a system of equations. In order to solve it, it is useful to define *a priori* two similar matrices F_d and F_r, without physical entity, whose elements would be as follows, (for a volume contained by three surfaces like the one depicted in figure 13):

$$F_d = \begin{pmatrix} F_{11} * \rho_1 & F_{12} * \rho_2 & F_{13} * \rho_3 \\ F_{21} * \rho_1 & F_{22} * \rho_2 & F_{23} * \rho_3 \\ F_{31} * \rho_1 & F_{32} * \rho_2 & 0 \end{pmatrix} \tag{23}$$

$$F_r = \begin{pmatrix} 1 & -F_{12} * \rho_2 & -F_{13} * \rho_3 \\ -F_{21} * \rho_1 & 1 & -F_{23} * \rho_3 \\ -F_{31} * \rho_1 & -F_{32} * \rho_2 & 1 \end{pmatrix} \tag{24}$$

Where F_{ij} are the radiative exchange fractions or factors from surface i to surface j found previously, and ρ_i is a new term, defined as the coefficient of diffuse or direct reflection which can be attributed to surface i. It can be obtained as the quotient between the energy received and the energy effectively emitted.

This is the reason why in matrix F_d described at 23, the element in column 3, row 3, is zero as a planar disk does not send radiation to itself.

However, if we should have a curved surface it would be necessary to substitute the former element by $F_{33}*\rho_3$. On the contrary, in a volume defined by planes, all the elements in the diagonal of matrix F_d are equal to zero.

Once the value of these matrices is obtained, it is easy to establish a relationship between direct and reflected radiation:

$$F_r * E_{ref} = F_d * E_{dir} \tag{25}$$

$$F_{rd} = F_r^{-1} * F_d \tag{26}$$

$$E_{ref} = F_{rd} * E_{dir} \tag{27}$$

As the value of reflected radiation is known, the problem is solved. However, we have to bear in mind that the minimum of surfaces in an actual room would be six.

Recently, the author has developed software for up to twelve surfaces inside a room. This will augment precision at the cost of a lengthier input of data.

The process of interreflection can be repeated many times until no significant changes in reflected radiation are observed.

Again, a simple case of this reiterative process occurs in the sphere and is often used as a substitute for the calculations described.

In equation 10 and successive, it was stated that the energy received by a point in the sphere from any surface contained in the same sphere was equivalent to the ratio between the area of the emitting surface and the total area $4*\pi*R^2$, this can be written W/A.

After infinite rebounds, the energy reflected on the sphere would be:

$$E_{ref} = E * \frac{W}{A} * (\rho + \rho^2 + \cdots \rho^n) \tag{28}$$

As,

$$\lim_{n \to \infty} \left(\frac{\rho^{n+1}-1}{\rho-1} - 1 \right) = \frac{\rho}{1-\rho} \tag{29}$$

$$E_{ref} = E * \frac{W}{A} * \left(\frac{\rho}{1-\rho} \right) \tag{30}$$

ρ in the former equations means the average of the reflection coefficients ρ_i and E is the direct energy emitted by the source. Therefore, this equation is suitable for all kinds of volumes, but its accuracy diminishes as the actual room is dissimilar to a sphere. Under these circumstances, equation 27 is preferable to equation 30.

As the reflectance of the room surfaces may be shifted at will, it is possible to conceive some of great absorptiveness which can be equated to virtual or transparent surfaces and, in that manner we would deal with semi-open spaces or non-reflecting elements. For instance, the former enables us to address the urban canyon and consequently the radiative transfer that takes place in urban spaces. To perform that kind of analysis, the energy exchanges in parallelepipeds must be obtained in advance.

4. Radiative transfer between plane surfaces

There is no special problem in extending what we had found for the vertical component of the disk in equation 15 and the corresponding section. Only a cartesian reticule instead of polar is needed.

This new equation, albeit not difficult, especially for the computer, is greatly altered, when compared with the simple $h/2*R$.

$$f(x,y) = \frac{E}{2} \left[1 - \frac{x^2+y^2+z^2-a^2}{\sqrt{(x^2+y^2)^2 + 2*(x^2+y^2)*(z^2-a^2)+(z^2+a^2)^2}} \right] \tag{31}$$

In expression 31, z is the vertical distance to the point considered on the graticule and a, is the radius of the source as before. The resulting field for the horizontal direction can be seen in figure 14.

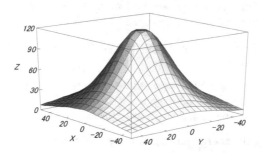

Fig. 14. Radiative field on a horizontal cartesian surface due to a disk source

If the disk turns to a rectangle, then the factor that gives the energy fractions, shows an entirely different configuration,

$$f(x,y) = \frac{E}{2}\left[\frac{y}{\sqrt{z_2^2+y}}\left(arctan\frac{x+\frac{a}{2}}{\sqrt{z_2^2+y}} - arctan\frac{x-\frac{a}{2}}{\sqrt{z_2^2+y}}\right) - \frac{y}{\sqrt{z_1^2+y}}\left(arctan\frac{x+\frac{a}{2}}{\sqrt{z_1^2+y}} - arctan\frac{x-\frac{a}{2}}{\sqrt{z_1^2+y}}\right)\right] \tag{32}$$

Where a, is the width of the rectangle and the height is z_2-z_1.

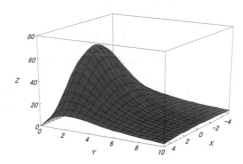

Fig. 15. Radiative field on a horizontal cartesian surface due to a vertical rectangular emitter

As in the case of the circular source, the horizontal component of the radiation vector has a more complex formulation. The reader may find further information in the references given. Without this component and a third perpendicular one, because now the analysis is performed in cartesian space instead of spherical, the vector remains undefined. This may lead to considerable inexactitude.

Turning to the exchange between surfaces as a whole, the mathematical procedure is indeed cumbersome and exceeds the scope of this chapter, though the author has been capable of completing it (see references). In order to program the solution for the computer in a more efficient way, it is often convenient to follow the method proposed by Yamauchi[5] (Yamauchi, 1934).

[5] Paradoxically, this clever procedure was invented to overcome the lack of computational capacities.

In this procedure the value of function Φ, is found for a different set of angles related with the dimensions of the surfaces considered.

$$\emptyset_0(\mu) = \frac{1}{2} * (\mu - \frac{1}{2} * \tan(\mu) * \ln(\sin(\mu)) + \frac{1}{2} * \cotan(\mu) * \ln(\cos(\mu))) \tag{33}$$

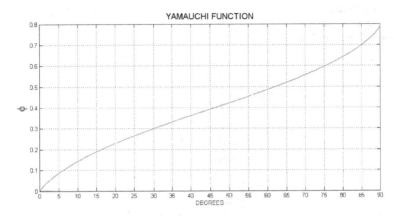

Fig. 16. Graph of the Yamauchi function from 0 to 90 degrees

The set of angles previously mentioned is obtained by looking at figure 17 and equations 34 to 36.

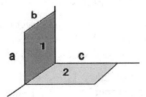

Fig. 17. Perpendicular surfaces with a common edge used to find the energy transfer

$$\propto = arctan(b/a) \tag{34}$$

$$\beta = arctan(b/c) \tag{35}$$

$$\beta_1 = arctan\left(\frac{b}{\sqrt{a^2+c^2}}\right) \tag{36}$$

Thus, the fraction of energy from surface 1 to surface 2 is,

$$F_{12} = \frac{2}{a*\pi}[a * \emptyset_0(\alpha) + c * \emptyset_0(\beta) - \sqrt{a^2 + c^2} * \emptyset_0(\beta_1)] \tag{37}$$

5. Radiative transfer in a space composed of several volumes

A further step that the author has devised consists of subdividing a complex space in several simpler volumes whose performance is eventually concatenated. The author has called this procedure the Superposition Principle of Radiation.

This situation is convenient for the treatment of several building features that perform as radiation filters such as canopies, awnings, louvers and even courtyards or reflective ponds.

For instance, in a system of louvers it is possible to isolate the volume seen in figure 18 and treat it as a single space with three virtual faces. Once the radiation that reaches the surface of the glass is obtained, the procedure is the same as for a room without louvers but the emissive power used for the window is the previous value and not the one applicable for the unobstructed orientation.

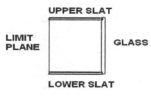

Fig. 18. Subdivision of louvers in a protected window

In this way, most of the problems derived from the geometry of the design are solved and radiation filters can be properly evaluated. Previously they were only considered obstructions without any potential to add for the energy balance.

6. Discussion and simulation examples

With all that we have expounded thus far, the reader is in the position to extract useful consequences to find the performance of radiation in buildings, either existing or projected.

Using his original software, based on the former section, the author has conducted a wide variety of simulations around the world. Most of them were validated by means of direct monitoring, both automatic and manual where available.

However, some provisos have to be taken into account. First of all, it has been assumed that radiation is emitted in a diffuse manner following Lambert's law. While this may be true for many materials especially modern ones, when dealing with heritage buildings such properties may not be accurate. In fact, the reflectance of surfaces at no longer extant architectural spaces remains largely unknown.

Even more difficult is the question of glazing in ancient buildings. Transparent glass panes which follow quantum dynamics in transmitting radiation are relatively modern. An added constraint is the fact that, recently, a wide variety of systems capable of selective or holographic transmission has been made accessible to designers and builders.

The main solution has been to define a directional or volumetric transmittance for glazing. This is a similar concept to the well-known photometric curve and gives us the spatial or spectral properties of glass-emitters. For the time being, these transformations can only admit bi-ellipsoidal form in the author's software.

For these and other reasons, interferences of radiation like diffraction and scattering, though predictable, are not handled in their entirety. Fortunately these phenomena are not very common in the building industry, especially because they may lead to visual discomfort and are generally avoided by users and designers.

Once the radiative transfer is settled for a given space through its geometric and optic features, the amount of renewable energy available is known. This may become an important figure in the energy analysis or may have a thermal or visual correlate. The visual results are more intuitive than the thermal ones.

To find the temperature field due to radiation on a surface, Stefan-Bolzmann's law has to be invoked and significant differences with the luminous domain emerge. The first and more relevant one is that the temperature of the surfaces considered has to be found or estimated since there are no elements at 0 K in buildings. The author's and other correlations help in this respect but may not be definitive. A second proviso is that reflectances for thermal radiation are not similar to those in the luminous domain. Fortunately, most of them fall into the range of 0.9 for interior building surfaces.

Finally, if due to ventilation a convective field coincides with that provided by radiation, the latter, according to our experiments, will not be significantly altered in the short term because what is mostly affected is the thermal sensation.

With all the former in mind, the author would like to present the simulation cases of two paradigms of ancient Roman architecture, whose accurate radiative performance was largely unknown: the Pantheon and its superb baroque evolution the Church of Sant'Andrea all Quirinale. The architect and sculptor of light Gian Lorenzo Bernini completed this masterpiece, considered to be his own spiritual retreat.

Following the discussion of radiation in centralised spaces, a building currently under construction, the new railway station at the airport of Barcelona (Spain) is briefly presented in an effort to show how simulation can help in the design process and assessment.

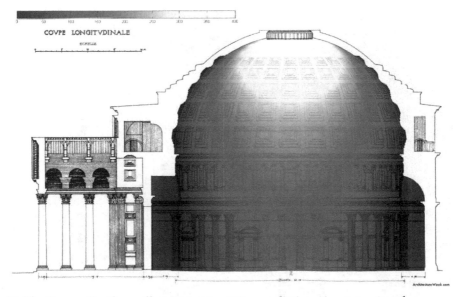

Fig. 19. The Roman Pantheon illuminated by diffuse radiation of an intensity of 10000 lumen/m² (lux). A typical situation in autumn and spring. Scale 0 to 400 lux

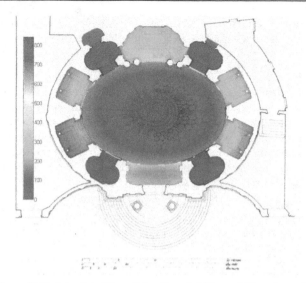

Fig. 20. Sant'Andrea all Quirinale's Church by Bernini (Rome) illuminated by direct solar radiation in winter. Values in lux(0-800)

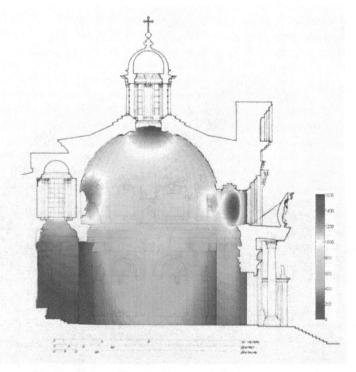

Fig. 21. Sant'Andrea all Quirinale's Church. Section under direct solar radiation in winter. Values in lux(0-1600)

The final case to be introduced is the Rautatalo building of 1955, by the modern Finnish master Alvar Aalto. Originally a department store, it beckoned Helsinki's citizens by its intelligent use of luminous radiation, enhanced by conical skylights subtly adapted to the solar path in this northern city.

In the first two examples, luminous radiation is nuanced and constant for the lower spaces. It is outlined that the values for the Pantheon were not significant (sometimes, under 200 lux) and this fact may have led to the introduction of vertical windows in the drum of the cylinder by late Renaissance or Baroque epochs. Radiative performance does not show an acute seasonal variation, but allows for sunshine to reveal certain decorative details of the structure adding to the reputation of spiritual luminous atmosphere that encompass the work of Bernini. A hall of more than 300 square metres featuring a consistent level of 800 lux with only nine carved windows and the magnificent lantern is remarkable for the 17th century.

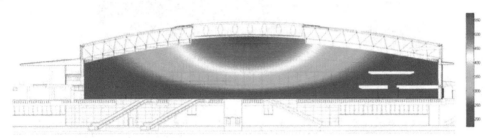

Fig. 22. Section of the new railway station in Barcelona. Radiation design by the author. Project by the architects Cesar Portela and Antonio Barrionuevo. Values in lux (0-600)

Changing the scale for the modern requirements of transportation spaces which have become the cathedrals of our time, the author proposes a lighting design in which the oculus reaches a diameter of 35 metres and the radiative energy is distributed by means of massive aluminium louvers with a height exceeding 3 metres in total. The simulation shows good levels and an acceptable raise of temperature at the glazed aperture due to the mild climate of Barcelona.

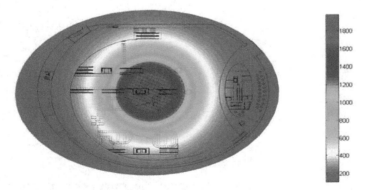

Fig. 23. Plan of the railway station in summer. Values in lux

The last example, the Rautatalo building, brings the reader back to the efforts of the modern movement in architecture to control radiation. With 40 skylights it was subsequently adapted to many projects around the world, which generally speaking fared less well than the original for climatic and economic circumstances.

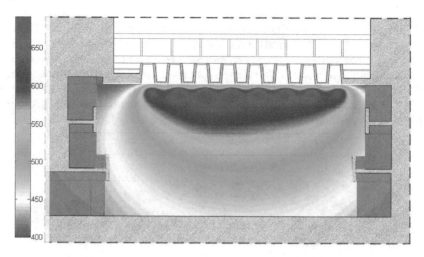

Fig. 24. The Rautatalo building of 1955 by Alvar Aalto, Helsinki. Simulation of 40 skylights (8*5), performed in June with direct sunlight and monitored on 21st of June 2011. Values in lux

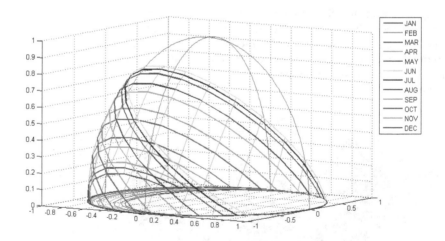

Fig. 25. Solar Chart of Helsinki. Latitude 61.16 degrees north

This climate-responsive building would remind the reader that, in order to produce universal results there is the need to consider local weather parameters.

7. Weather data

As figure 25 tries to evoke, weather data for radiation must be based on sunlight availability. At the time of writing this chapter, many measurements of horizontal irradiation in the world have been recorded and also correlations for vertical surfaces of different orientations are available. The author recommends the following based on altitude θ and azimuth φ from the south direction (in radians).

$$E_v = 4000 * \theta^{1.3} + 12000 * sin^{0.3}\theta * cos^{1.3}\theta * [(2 + cos\varphi)/(3 - cos\varphi)]) \qquad (38)$$

The author's software is capable of combining these results with sunlight probability for any location in the world to obtain annual, monthly and hourly distribution of irradiance on vertical and horizontal surfaces. (See tables 1 and 2).

	NORTH	EAST	SOUTH	WEST	HORIZONTAL
DIFFUSE	46,57	67,04	85,19	67,04	113.79
DIRECT	39,40	283,89	271.16	283,89	290.31
GLOBAL	85,97	350,93	356.35	350,93	404.11

Table 1. Mean annual radiation (W/m²) by orientation in Rome. Italy.
Latitude= 43.41° North

	SOUTH	EAST	NORTH	WEST	HORIZONTAL
DIFFUSE	72,56	81,02	77,41	81,02	152,44
DIRECT	147,85	348,94	173,78	348,94	428,05
GLOBAL	220,42	429,96	251,19	429,96	580,49

Table 2. Mean annual radiation (W/m²) by orientation in Quito. Ecuador.
Latitude= 0.3° South

Nonetheless, the former data are averages and not a substitute for measured registers and much less for instant values. Only simultaneous monitoring at intelligent buildings can achieve real time input in our simulation model.

What is recommended in this chapter is a likely figure intended to handle the problem with reasonable accuracy in the frequent absence of more detailed information. This will help to design building features that save energy and comply with the most relevant weather conditions at each climate[6].

---

[6] Klymax κλίμάξ in Greek from where the word climate derives, means "stairway".

8. Conclusion

The author has produced some innovative tools which prove to be highly efficient and compatible with those currently employed in architectural and engineering projects. In doing so, it is his firm belief that creativity and freedom of design in the realm of solar radiation will be much enhanced. The so-created software is universal and it aims to bridge the gap in solar design between developed and non-developed regions.

The conscious application of these techniques also brings new possibilities to benefit from solar radiation in our own homes. This is in the author's hope, a good way to help the peoples in the world, in a moment of turbulence and social unrest.

The main drawback found is the lack of preparedness of many architects and authorities to implement these methods in the decision-making process.

With the warp thus created fabrics can be woven as usual, although it is also possible to make a sort of net out of it and lie down at ease, or perchance play with it as a makeshift harp and see how it sounds! For as Coleridge once wrote[7],

And what if all of animated nature

Be but organic Harps diversely fram'd

That tremble into thought

9. Acknowledgment

The author wants to recognize the help of the following people: Miss Patricia Karlsson of the National Pension Institution of Helsinki, who was instrumental in revealing Aalto's mastery of light; the kindness and assistance of Professor Mojita Navvab of the University of Michigan are deeply appreciated. Among many other favours, he granted access to MacAllister's original papers. Karma, my research group of the University of Seville was extremely supportive, especially Professors Jesus Pulido and Viggo Castilla; and in Japan, Professor Tetsushi Okumura of the University of Nagoya has always been helpful.

Finally, it is the author's wish to praise the Japanese people for their tenacity and profound respect of Nature, regardless of the tragic blows that She may inflict on this beloved land.

10. References

Ashdown I. (2004). *Radiosity: A Programmer's Perspective*. John Wiley & Sons Inc. New York, 1994. Available from http://www.helios32.com.
Baker, N. V., Fanchiotti, A., Steemers, K. N.(1993). *Daylighting in Architecture. A European Reference Book*. Commission of the European Communities. Directorate General XII.

---

[7] Samuel T. Coleridge. *The Aeolian Harp*. 1795

Cabeza Lainez, J.M. (2010). *Fundamentos de Transferencia Radiante Luminosa.* (Including software for simulation). Netbiblo. Spain.

Cabeza Lainez, J.M. (1999). Scientific designs of sky-lights. *Conference on passive and low energy architecture* (PLEA). Brisbane. Australia.

Cabeza Lainez, J. M. (2007). The Japanese experience of environmental architecture through the works of Bruno Taut and Antonin Raymond. Vol.6 I. Pgs. 33-40. *Journal of Asian Architecture and Building Engineering* (JAABE). ISSN: 1346-7581

Cabeza Lainez, J. M. (2007).Radiative performance of louvres, Simulation and examples in Asian Architecture. *IAQVEC.* Volume III . ISBN: 978-4-86163-072-9 C3052 \4762E. Sendai (Japan).

Cabeza Lainez, J. M. (2008). The Quest for Light in Indian Architectural Heritage. Vol.7 I. Pgs. 17-25. *Journal of Asian Architecture and Building Engineering* (JAABE). ISSN: 1346-7581.

Cabeza Lainez, J.M. (2009) *Lighting Features in Japanese Traditional Architecture.* In "Lessons from Traditional Architecture". Editors, Yannas, S., Weber, W. Earthscan. London. ISBN 9781844076000

DiLaura D. L. (1999). New procedures for Calculating Diffuse and Non-Diffuse Radiative Exchange Form Factors. *Proceedings of ASME.*

Feynman, R. P. (1990). *Quantum electrodynamics: The Strange Theory of Light and Matter.* Penguin Books.

Fock, V. (1924). *Zur Berechnung der Beleuchtungsstärke.* Optisches Institut St. Petersburg.

Garibaldi, C. (1994). *Simulazione deterministica della radiazione solare nella chiesa di San Lorenzo a Torino.* CNR Italy.

Higbie, H. H. (1934)*Lighting Calculations.* John Wiley and Sons. New York.

Holman, J.P. (1997) *Heat Transfer.* Mac Graw-Hill. New York.

Hopkinson, R. G.; Petherbridge, P.; Longmore, J. (1966) *Daylighting.* London. Heinemann.

Kimura, K. (1977). *Scientific Basis of Air Conditioning.* Amsterdam. Elsevier.

Lambert J. H. (1764).*Photometria. sive de mensura et gradibus Luminis, Colorum et Umbrae.* Editor. D. DiLaura. IESNA. 2001.

MacAllister, A. S. (1910). Graphical Solutions of Problems Involving Plane-Surface Lighting Sources . *Lighting World* 56. No.1356.

Moon, P. H; Spencer D. E. (1981) *The Photic Field.* The MIT Press. Cambridge. Massachusetts.

Moon, P. H. (1962)*The Scientific Basis of Illuminating Engineering.* Dover Publications. New York.

Moore, F. (1991). *Concepts and Practice of Architectural Daylighting.* Van Nostrand Reinhold. New York.

Ne'eman, E. (1974) Visual Aspects of Sunlight in Buildings. *Lighting Research and Technology.* Vol 6. N° 3.

Pierpoint, W. (1983). A Simple Sky Model for Daylighting Calculations. *International Daylighting Conference.* Phoenix.

Robbins, C. L. (1986). *Daylighting. Design and Analysis.* Van Nostrand Reinhold. New York.

Shukuya M, (1993) *Hikari to Netsu no Kenchiku Kankyôgaku* –Light and temperature in Environmental Science - (in Japanese). Maruzen. Tokyo.

Yamauchi, J. (1927). The Light Flux Distribution of a System of Inter-reflecting Surfaces. *Researches of the Electro-technical Laboratory*. No. 190.Tokyo. (In Japanese).

Yamauchi, J. (1929). The Amount of Flux Incident to Rectangular Floor through Rectangular Windows. *Researches of the Electro-technical Laboratory*. No. 250.Tokyo.

Yamauchi, J. (1932). Theory of Field of Illumination. *Researches of the Electro-technical Laboratory*. Tokyo. No. 339.

2

Innovative Devices for Daylighting and Natural Ventilation in Architecture

Oreste Boccia, Fabrizio Chella and Paolo Zazzini

D.S.S.A.R.R., University "G. D'Annunzio"Chieti, Pescara

Italy

1. Introduction

It is known that every human activity is better carried out in the presence of natural light than in the absence of it. This is probably due to the comfortable feeling of the occupants respect to the perception of the flowing of time, which is impossible when artificial lamps are the only light sources in the environment. In many cases the natural light is absent, as in underground areas of a building, or insufficient, as in the case of large plant area rooms, where the windows on the perimeter walls are not able to illuminate the whole environment being too small or too distant from the centre of the room. Moreover the indiscriminate use of electric light even if the daylight is available, due to a deplorable habit of the occupants of the workplaces, is a fairly common practice and it increases the energy consumption in buildings that already covers about 40% of the total energy consumed worldwide.

Many technological devices have been developed with the aim of contributing to an efficacy energy saving by using daylight in buildings, such the light pipes, that are able to transport the natural light away from the collection point, usually on the roof-top of the building, for example in hypogeum environments. There are two different types of light pipes depending on whether they are equipped with fixed or mobile collector. In recent years the authors carried out an intense experimental and numerical analysis with the aim to evaluate the daylight performances obtained by light pipes equipped with fixed collectors (Chella et al. 2006, Baroncini et al. 2006, Zazzini et al. 2006). Moving from the experience gained in the work, the authors developed some innovative devices which improve the performances of the traditional light pipes. The first one is called "Double Light Pipe" (DLP) and it is able to transport daylight into a two floors underground building. It is an evolution of the traditional light pipe. The second one is a further evolution of the DLP, which allows transport of daylight into underground areas of a building as a DLP, and guarantees the necessary change of air by natural ventilation. It is named "Ventilated Double Light Pipe" (VDLP) and it is simultaneously able to introduce daylight and fresh air in underground areas of buildings or rooms without direct interface with outdoor. This device was recently presented at the World Renewable Energy Congress 2011 (Boccia et al. 2011). Through a numerical analysis, the performances obtained by the VDLP and its applicability in architecture are examined. The numerical data allow the examination of the possibility of generating an architectonical space from the availability of daylight and fresh air by natural ventilation.

The main goal of this work is to propose an architectonical space equipped with a certain number of VDLP, in which the geometry is generated by the availability of daylight and the possibility of effecting an efficacy of natural ventilation by the VDLP.

2. Traditional light pipes

The lack of daylight in buildings is often a cause of large amount of electric energy consumption. This problem is particularly present in the underground buildings or in industrial or commercial large plant area edifices in which electric light is used all the time in the spaces occupied. In recent years many technological devices, called light pipes, sky lights or tubular sky lights, have been proposed in order to offer a solution to the problem. They are able to collect natural light with a mobile or fixed collector and redirect it into the interior spaces away from the collection point.

Light pipes with mobile collectors can be equipped by heliostats that are solar tracking devices able to rotate with respect to one or two axis, so optimizing the direct solar radiation inlet into the tube.

Fixed collectors are cheaper than mobile collectors. They are usually constituted by a transparent polycarbonate dome with the function of collecting natural light from the sun and the sky and redirect it into the tube. In many cases they also have Fresnel lenses able to concentrate direct solar radiations coming from various directions in the direction parallel to the pipe.

The entrance point is in most cases located on the roof-top of the building or alternatively on one of the external walls. A free collection of light is needed for optimizing the efficiency of the device so that any obstruction of solar radiations from adjacent structures may be avoided.

The tube is internally covered by a highly reflecting film along its length from the collector to the diffuser for long-distance transport of natural light in the underground areas or spaces without any direct interface to outdoor. The larger the diameter of the collector, the longer the distance that can be covered by light with minimum dissipation of energy.

Light pipes on the market have diameters of the collector varying between 0,25 to 1 m. The bigger light pipes are usually adopted in commercial or industrial applications. The smaller ones are used for residential installations and they are typically 1-5 m long.

Natural light is more efficiently transported with straight pipes, but curved or angled tubes can be adopted if necessary. Light transmission losses are reduced by the highly reflecting film applied on the internal surface of the tube. Reflectivity ranges from 0,98 to 0,995.

In recent years many authors carried out numerical and experimental analysis with the aim to determine the performances of light pipes (Carter 2002, Zhang et al. 2002, Jenkins et al. 2003) and theoretical or empirical calculation methods were set up by various researchers for the prediction of the illuminance distribution inside a room equipped by light pipes (Jenkis & Muneer 2004, Jenkis & Zhang 2004), but standard design methods have not yet been developed. The difficulty in standardizing the calculation methods is mainly connected with the influence on internal illuminance distribution of the temporary weather variations that commonly occur during a day, a season or a year, and it is very difficult to describe the real climatic situation with a simplified model based on standard conditions that are necessary for a correct design of the system.

And finally in a room of a building equipped by light pipes, a polycarbonate light diffuser is applied to the ceiling in order to spread light into the room.

The authors published in Baroncini et al. (2006) the results of an experimental comparison between two types of commercial diffusers installed on the same light pipe for residential applications, and demonstrated that the geometry adopted in the diffuser can be modified for spatial distribution of light.

If compared to traditional skylights, the light pipes offer better performances in term of summer heat gain reduction, due to a less visual contact with outside. In addition the availability of natural light in hypogeal environments improves the occupants' well-being, avoiding over illumination effects. A significant energy saving contribution is offered by light pipes installations, by reducing the electric energy consumption.

Light pipes are often characterized by considerable dimensions, their installation in the centre of a passage room is unacceptable, being very bulky systems. When a light pipe is installed to transport natural light from the rooftop of the building to an underground room, the passage spaces are occupied by a very voluminous system which constitutes an undesirable stumbling block and makes it inconvenient to install it in the centre of the room.

3. The Double Light Pipe

The Double Light Pipe (DLP) is an innovative device developed by the authors to illuminate a two - level underground room. It is an improvement of the traditional light pipe, moving from the idea that the installation of a traditional light pipe in the centre of a two - floor building, with the aim of illuminating the lower underground spaces, is impossible because of the encumbrance of the device.

Since the DLP is able to distribute daylight both in to the passage and the final room, it is proposed to be a solution for the problem. In fact, if a double light pipe is used instead of a traditional one, it is in good agreement with the architectural principle because it illuminates the passage room, so justifying its bulkiness.

The DLP is an improvement on the traditional light pipe. It consists of two concentric tubes, the internal one which illuminates the final room like the traditional light pipe. The same reflective film (ρ=99,5 %) covers both the internal and the external surfaces of the inner pipe while the second one, concentric to the first, is made of a transparent material such as polycarbonate and it is installed so as to create a hole between the two pipes. It allows transmission of a portion of daylight, captured by the collector on the rooftop, into the interior spaces crossed by the system.

In figure 1 a picture is shown of a reduced scale (1:2) prototype of the DLP set up by the authors in the laboratory of Technical Physics of the University "G. D'Annunzio" of Pescara and the concept on which the device is based.

In order to diffuse light in the passage room, avoiding too intense reflections from the interior pipe, a thin plastic diffusing film is applied on the interior surface of the transparent external pipe. It is characterized by very precise 90° micro-prisms on one side and is made smooth on the other so that it may distribute light in a more diffuse way in the crossed room.

Thanks to its optical properties, it can either reflect or transmit, depending on the angle of incidence of the light. radiation. Light is reflected if the angle is less than about 27° with respect to the axis of the prisms, and transmitted if the angle is greater than 27°.

The diffusing film may be applied on the upper portion of the pipe (30-50 cm) in order to avoid glare. Figure 2 shows the construction steps of the system and its illuminating function in the passage and final room.

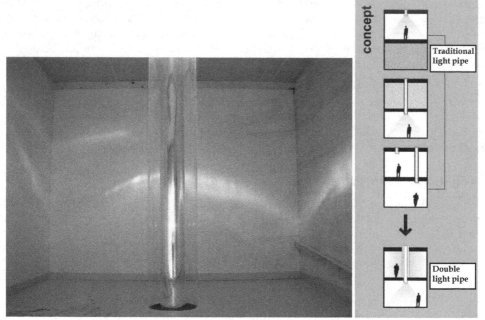

Fig. 1. Reduced scale prototype of a DLP and the concept of the system

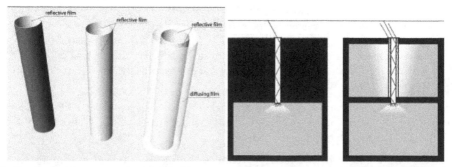

Fig. 2. Evolution from a traditional light pipe to a double light pipe and its lighting function in a two floor underground room

Since the DLP provides light in the room, it can be installed in the centre of the area. The technological design of the DLP and data about its performances are shown in Chella et al. (2007), Baroncini et al. (2008), Baroncini et al. (2009), Baroncini et al. (2010).

The DLP can be installed either in underground rooms or large plant area spaces in which the windows are placed on the perimeter walls and they are not able to guarantee an efficient distribution of natural light in the whole environment. In such cases, the DLP is very suitable, particularly if a soft light distribution is required, characterized by a certain degree of uniformity, such as in exhibition rooms, museums and the like, in which direct solar radiations must be avoided because of the risk of glare and deterioration of the artworks sensitive to light. For this reason, in museums or other similar places, even if not in hypogeal rooms, any direct visual contact with outdoor is usually avoided and artificial light is used all the time in such environment.

The collector of a DLP is larger than a traditional one. It is able to collect light and redirect it both to the inner pipe and to the hole between the two pipes. It can be a dome or plane cover made of transparent material such as polycarbonate, equipped by Fresnel lenses or not.

In the reduced scale prototype set up by the authors, the simplest type of collector was adopted, a plane transparent polycarbonate device placed on the rooftop, with the intention of evaluating the performances of the transmitting and diffusing devices in the least favourable conditions. Any enhancement of shape and materials, with the aim of optimizing the possibility of collecting daylight, will obviously increase the performances of the DLP.

The inner pipe consists of an aluminium sheet folded like a tube which is covered by a multilayer highly reflective film both over its internal and external surfaces. Due to its reflecting characteristics, very efficient multiple reflections take place in the pipe; direct and diffuse natural light are channelled downward to the diffusion point in the final room.

The outer pipe, concentric to the inner one, is made of a transparent polycarbonate tube, internally covered partially or completely on its surface by a diffusing material, which distributes light into the passage spaces.

At the end point of the light path, in the final room, a conventional polycarbonate diffuser with a regular prismatic geometry was adopted similar to the traditional light pipe. In all respects the inner pipe performs like a traditional light pipe.

Figure 3 shows a two-dimensional section of a two - level underground building equipped by a DLP, in which the used materials are evident, while in Figure 4 an example of application of a DLP to an underground museum is shown. [Chella et al. 2007]

3.1 Experimental analysis on the Double Light Pipe

The authors carried out an intensive experimental and numerical analysis on the DLP. The experimental analysis was carried out on reduced scale models of the system, while for the numerical analysis, Radiance and Ecotect were used, two reliable soft-wares commonly used for daylight simulations (Chella et al. 2007, Baroncini et al. 2008, Baroncini et al. 2009, Baroncini et al. 2010) .

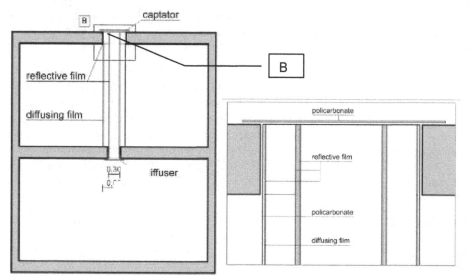

Fig. 3. Two dimensional section of a two levels building equipped by a DLP

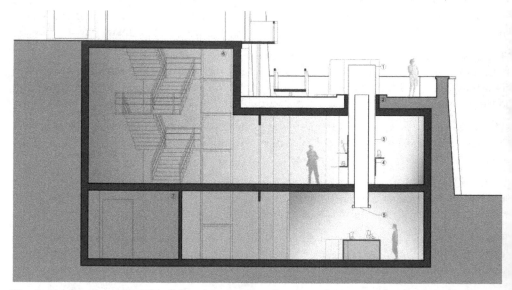

Fig. 4. Application of Double light pipes in a two levels underground museum.

The experimental data measured on a 1:2 scale DLP in three different tests are shown in the following figures. The results report the illuminance distribution in the passage area measured in 12 internal positions on a horizontal work-plane 40 cm high above the floor and outside of the building close to the collector on the roof-top as well are presented in the following figure.

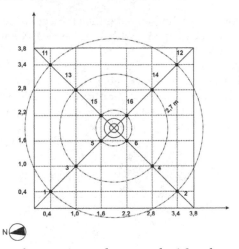

Fig. 5. Measure positions in the experimental tests on the 1:2 scale model of the DLP

Figure 6 shows the results of a test carried out under intermediate cloudy sky between 9 am and 1 pm. The external illuminance ranged between 17 and 76 klux, with a mean value of about 40 klux. It is evident that the interior illuminance trend is very similar to the external one in every position in the absence of direct solar radiation. Some peak values of illuminance observed in positions 1 and 2 close to the corners of the room at about 11 am may probably due to particularly intense reflections from the device. The ratio between E_{in} and E_{ext} is less than 1 % all over the work-plane.

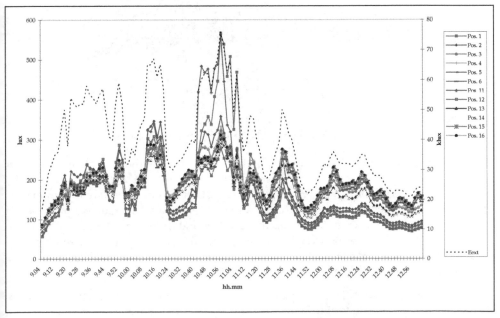

Fig. 6. Experimental data on a 1:2 scale model of a DLP under cloudy sky

In Figure 7 data are shown regarding an experimental test effected on the same 1:2 scale model of the DLP with intermediate sky with sun, between 9 am and 4 pm, the external illuminance ranging between 18 and 73 klux with a mean value of about 56 klux. In this case, the peaks on the corner of the room are more evident than in the previous test, not only in positions 1 and 2, but also in positions 11 and 12 on the opposite side of the room, probably due to a more significant influence of direct solar radiations. Peaks of illuminance take place at points determined by the hourly positions of the sun in the sky and by the directions of high reflections generated by the system.

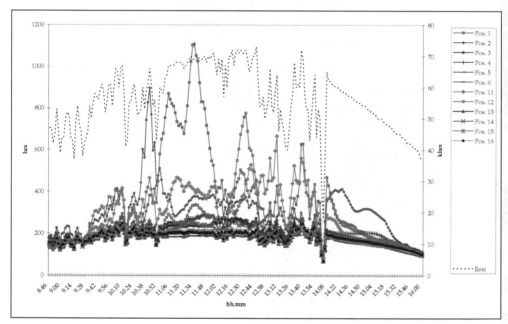

Fig. 7. Experimental data on a 1:2 scale model of a DLP under intermediate sky with sun

Finally, in Figure 8, the results of a two - day test on the same reduced scale model under clear sky with sun are shown. The test was carried out between 9 am of the first day and 3 pm of the second day. Corresponding to a more regular trend of the external illuminance, which is characterized by more powerful direct solar radiations (E_{ext} ranges from 11 to 74 klux, with mean value of about 51 klux in the first day and from 11 to 72 klux with a mean value of 55 klux in the second day) the internal illuminance is characterized by very frequent illuminance peaks in various positions. The greater the external illuminance the more frequent the presence of the peaks.

Starting from this consideration, it is likely that high reflections from the DLP involve glare phenomena for the occupants of the environment. In such cases the use of the diffusing film described above can be an effective antidote to this possibility.

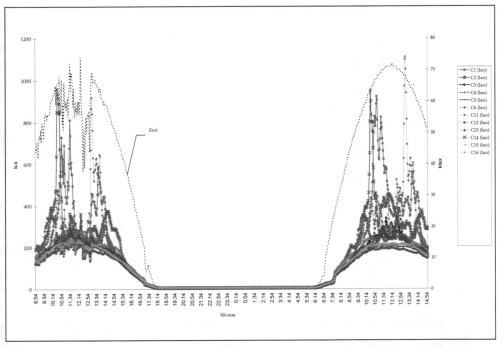

Fig. 8. Experimental data of a two - day test on a 1:2 scale model of a DLP under clear sky with sun

4. The Ventilated Double Light Pipe

People living or working in underground environments of buildings need both natural light and ventilation for well being. When air is made to flow by mechanical ventilation a considerable amount of electric energy has been consumed. Light pipes are often equipped by mechanical ventilation devices in order to assure comfortable optical and hygienic conditions for the occupants. For this need, the authors propose a modification of the DLP so that it can be used as an efficient tool for natural ventilation in the passage area in a two-level underground building, allowing a significant energy saving. This device is named Ventilated Double Light Pipe (VDLP).

The Ventilated Double Light Pipe (VDLP) is a further transformation of the DLP which geometry is modified to optimize the air coming in and out. To be more precise, in the VDLP, the inner tube is narrowed at the top, in order to generate a convergent section at the top to improve the air extraction, while the outer tube is narrowed at the bottom, so the hole between the two tubes has a convergent section downward for the air to come in. In this way each device can be alternatively used as an inlet or an outlet system. Particularly, in the inlet VDLP the inner pipe is closed at the top and the bottom, while at the outlet the air cavity between the two pipes is closed at the top and bottom, as shown in Figure 9.

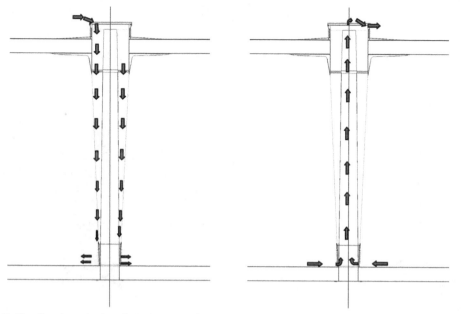

Fig. 9. Qualitative air circulation in an inlet and outlet VDLP

4.1 The modular square form equipped by four Ventilated Double Light Pipes

A certain number of VDLP can be installed in order to generate a modular architectural structure, by which more complex buildings of various geometrical forms may be obtained.

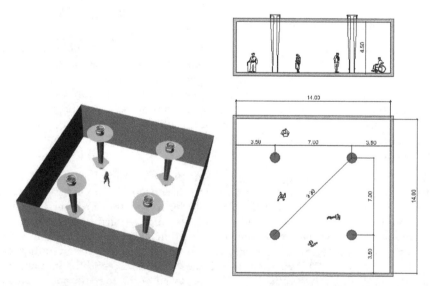

Fig. 10. Basic square modular form

In Boccia et al (2011), the authors proposed a two - level square basic modular form with four VDLP and presented the results of a numerical simulation, regarding daylight and natural ventilation performances. Each level consists of a 14x14 m² plant area room, 4,5 m high, in which two inlet and two outlet devices are installed, with the intention of introducing daylight in both the intermediate and the final room. At the same time natural ventilation by the VDLP installed is achieved in the passage room. Figure 10 shows the square model equipped by four VDLP.

4.2 Numerical analysis on the square modular form equipped with four VDLP

An intense numerical analysis was carried out by the authors in steady state condition with the aim of determining the daylight illuminance and luminance distribution by the sw Radiance and Ecotect, while the sw package Fluent / Airpak was used in order to calculate the air velocities and temperatures in the room, besides the thermal comfort index PMV and PPD.

4.2.1 Daylighting simulation

In Figures 11 and 12, the numerical illuminance distribution on June the 21st under Clear Sky, and on December the 21st under Overcast Sky respectively are shown, while in Figures 13 and 14 the internal luminance distribution in analogous conditions is shown. (Boccia et al. 2011)

Fig. 11. Internal Illuminance distribution on June the 21st – Clear Sky

Fig. 12. Internal Illuminance distribution on December the 21st – Overcast Sky

Fig. 13. Internal Luminance distribution on June the 21st – Clear Sky

Fig. 14. Internal Luminance distribution on December the 21st – Overcast Sky

In summer condition, under CLEAR SKY, an average value of about 800 lux is obtained in the environment with maximum values of about 2700 lux in positions close to the tubes. The illuminance spatial trend is not symmetric due to the influence of the reflections coming from the devices and coming also particularly from the intense direct solar radiations. In winter conditions, under OVERCAST SKY an average value of 150 lux is obtained with maximum illuminance values of about 500 lux. A similar situation is obtained regarding the asymmetric distribution of illuminance with peak values close to the VDLP. The images regarding the luminance distribution in the room (Figures 13 and 14) show how the luminances are all over comfortably with some areas on the walls that appear particularly bright with respect to the surrounding environment, as already evidenced in the picture of Figure 1. Only the upper portion of the tube is characterized by luminances so high that the risk of glare is present. In this case the adoption of a diffusing film applied on the internal surface of the transparent pipe is an appropriate antidote.

Starting from the results about daylighting the influence area of the VDLP has been determined in order to assure daylight comfort for the occupants. A centre distance between the VDLPs of 7 m has been chosen and on the basis of this geometric data the basic square modular form was designed.

4.2.2 Winter simulation on thermal comfort indexes

A further numerical analysis was carried out with the aim of determining the indoor thermo-hygrometric comfort index such as the air temperature and velocity, the PMV and PPD index, the mean age of air in winter and summer condition.

In winter condition the VDLP is not able to guarantee thermo-hygrometric comfort condition as confirmed by the air temperature data and the PMV and PPD comfort indexes (cfr. Figures 15 and 16).

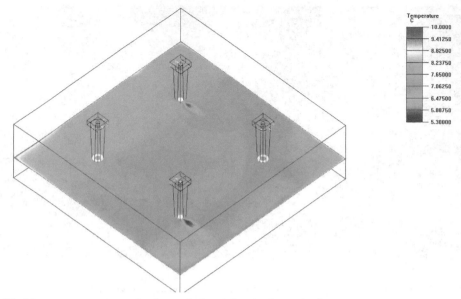

Fig. 15. Air temperature on a horizontal plan 1,7 m high on the floor

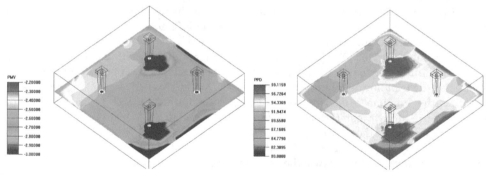

Fig. 16. PMV and PPD on a horizontal plane 1,7 m high above the floor

On the other hand, as underlined by graphs in Figures 14-17, the VDLP is able to guarantee a good natural air circulation. Figure 17 shows the air speed distribution on a horizontal plane 1,7 m above the floor in winter condition, with an external air temperature of 4 °C, in a two - level underground building, and a ground temperature of 15 °C. In the passage room the boundary walls and the floor are considered isotherm with a temperature of 15°C, while the ceiling is considered an adiabatic surface. The wind velocity corresponding to the inlet section of the collector is of 4 m/s. The wind direction is not considered because the device can vary its orientation in such a way that the inlet section is always normal to the wind direction.

Close to the inlet of the VDLPs, high values of air speed are observed (up to a maximum of about 0,5 m/s) while all over the environment the air velocity is less than 0,2 m/s with quite a uniform spatial trend. In Figure 18 the vertical air velocity at the centre of the room, with the same boundary conditions is considered. It is evident that the vertical trend of air speed is quite uniform with very comfortable values, except for positions near the floor in which the suction action of the device causes high values of air speed. The vertical sections shown in Figures 19 and 20, corresponding to the inlet and outlet of the VDLPs, indicate that they are able to effect an efficient air change in the environment as confirmed by Figure 21 in which data about the mean age of air are reported. It is clear that the in flow of fresh air and the out flow of exhaust air is efficiently effected by the devices so that the mean age of air is satisfactory all over in the environment particularly near the inlet tubes. In fact the calculated inlet flow rate is about 720 m³/h which correspond to 0,8 air exchange in an hour.

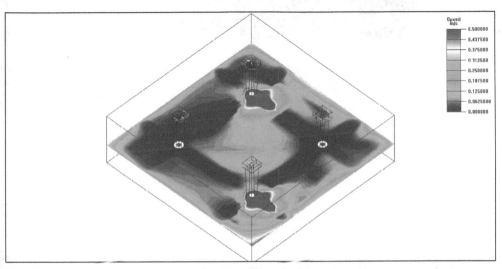

Fig. 17. Air velocity on a horizontal plane 1,7 m above the floor

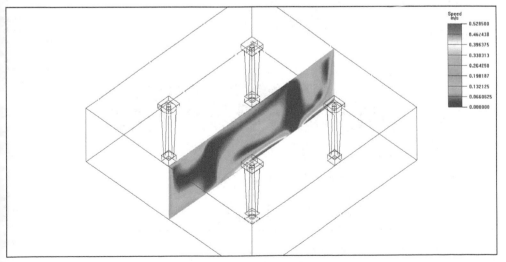

Fig. 18. Air velocity in the vertical plane at the centre of the room

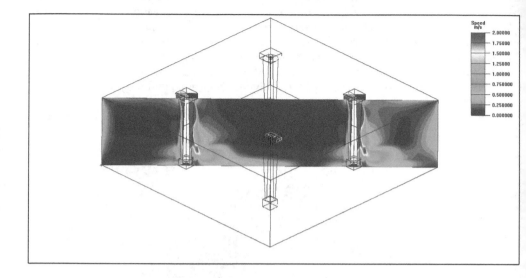

Fig. 19. Air velocity in a vertical section of the inlet tubes

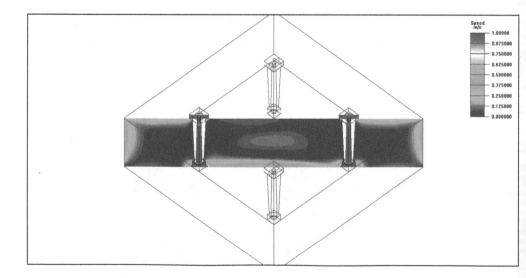

Fig. 20. Air velocity in the vertical section of the outlet tubes

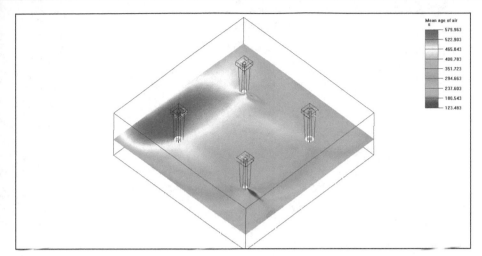

Fig. 21. Mean age of air on a horizontal plan 1,7 m high on the floor

4.2.3 Summer simulation of thermal comfort indexes

In summer condition, a better situation is observed in the thermal comfort indexes. The numerical simulation is carried out assuming an external air temperature of 30 °C, with a relative humidity of 70 % and a wind velocity of 4 m/s. The internal vertical walls and floor are considered isothermal with a surface temperature of 15 °C, while the ceiling is considered adiabatic. In addition the external horizontal surface of the collector is superheated with respect to the air and its surface temperature is assumed equal to 36°C due to the effect of solar radiations. As evidenced in Figures 22 and 23, the indoor air temperature is included between about 25 and 27 °C, both on a horizontal plane 1,7 m high above the floor and in a vertical plane in the centre of the room.

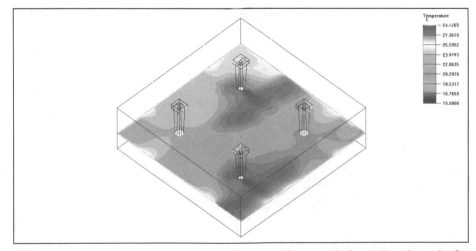

Fig. 22. Indoor air temperature in summer condition on a horizontal plane 1,7 m above the floor

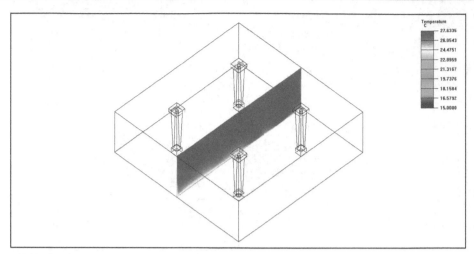

Fig. 23. Indoor air temperature in summer condition on a vertical plane in the centre of the room

Data about PMV and PPD indexes confirm that in the room a thermal comfort condition is obtained with the VDLPs in summer as shown by Figures 24 and 25, even though the internal relative humidity may not be satisfactory, ranging between 75 and 85 % all over the environment. It is an expected result since air is only refreshed without any dehumidification action, so the R. H. is increasing due to the decreasing temperature. Hence we can deduce that the internal air must be de-humidified in order to ensure complete satisfactory thermo-hygrometric indexes.

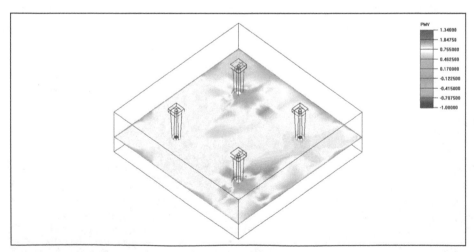

Fig. 24. PMV index on a horizontal plane 1,7 m above the floor

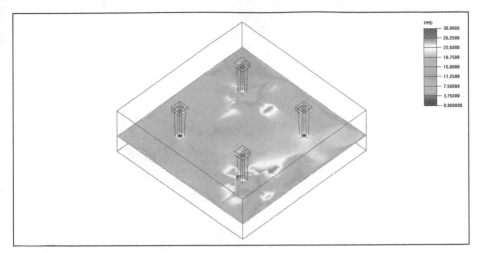

Fig. 25. PPD index on a horizontal plane 1,7 m above the floor

Regarding air velocity, we can say that the VDLPs seem to ensure good natural ventilation, since they are able to make available sufficient fresh air in the room and allow the necessary exchange of air for optimal hygienic condition. Figures 25-28 show the air speed spatial trend in the room on a horizontal plane and vertical plane, in the centre of the room and at the inlet and outlet devices. Being that in summer condition higher values of air velocity than in winter condition can be tolerated by the occupants of the environment, the air velocity is satisfactory both on a horizontal and vertical plane, and at the inlet, air flow-rate is able to sustain the necessary air exchange for hygienic comfort conditions. In fact the calculated inlet flow rate is about 540 m³/h, which correspond to 0,6 air exchange in an hour.

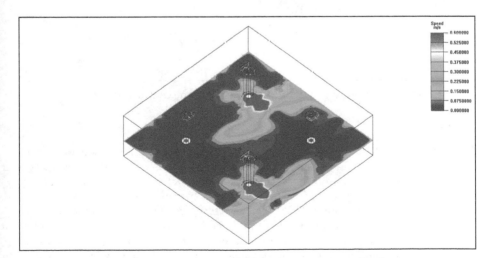

Fig. 26. Air velocity distribution on a horizontal plane 1,7 m above the floor

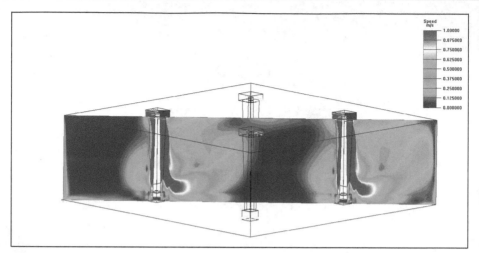

Fig. 27. Air velocity distribution on the vertical section at the inlet of VDLP

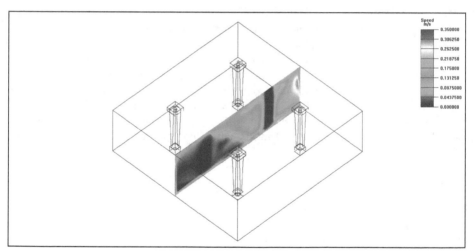

Fig. 28. Air velocity distribution on the vertical section at the centre of the room

5. Conclusions

In this work the authors show the results of an intense analysis carried out on some innovative devices for daylight transport and natural ventilation in underground areas of buildings or rooms without any direct connection with outside. Starting from the know-how, achieved in many years of numerical and experimental activity regarding the traditional light pipes, they propose two innovative devices: the Double Light Pipe (DLP) and the Ventilated Double Light Pipe (VDLP). By the DLP the problem of the encumbrance of a traditional light pipe installed in the centre of a two - level underground building is resolved because it illuminates both the passage and the final room, while the VDLP, in addition, is able to effect the necessary air exchange in the passage room.

The DLP was studied through a numerical and experimental analysis and its lighting performances are shown, while the VDLP was analysed by numerical methods by which its capacity to distribute fresh air in the passage room is investigated. Four VDLPs are composed in a basic modular form. The daylighting analysis allows the optimization of the centre distance between two adjacent devices to be set at 7 m. The successive numerical analysis regarding the thermo-hygrometric comfort indexes confirm that the square shape of a module, although perfectible, at this moment, is preferable to more complex geometric shapes because it offers the advantage of adapting to various possible compositions of modules.

In particular, the data of the numerical thermo-hygrometric and fluid-dynamics analysis enable us to conclude that the VDLP is an efficient tool for natural ventilation both in summer and winter conditions. Besides, in summer, it assures a certain degree of thermal comfort with regards to the spatial temperature in the environment as confirmed by data from PMV and PPD indexes. However it cannot effect the dehumidification action necessary to ensure complete thermo-hygrometric comfort conditions in summer. In winter, it cannot guarantee thermal comfort for the occupants of the passage room, but it can be considered an efficient tool for natural ventilation.

6. References

Baroncini, C. ; Boccia, O. ; Chella, F. & Zazzini, P. (2009). Double light pipe: experimental analysis on reduced scale models and comparison with numerical results, *LUXEUROPA 2009*, pp. 1041-1048, Istanbul, Turkey, September 07-11, 2009

Baroncini, C. ; Chella, F. & Zazzini, P. (2006). *Experimental analysis of tubular light pipes performances: influence of the diffuser on inside distribution of light*, 5th International Conference on Sustainable Energy Technologies SET 2006, pp. 219-224, Vicenza, Italy, August 30-September 01, 2006

Baroncini, C.; Boccia, O. ; Chella, F. & Zazzini, P. (2010). The Double Light Pipe, an innovative daylight technological device, *Solar Energy*, Vol. 84, No. 2 (February 2010), pp. 296-307, ISSN 0038-092X

Baroncini, C.; Chella, F. & Zazzini, P. (2007). Numerical and experimental analysis on Double Light Pipe, a new system for daylight distribution in interior spaces, *International Journal of Low Carbon Technologies*, Vol. 3, No. 2 (April 2008), pp. 110-125, ISSN 1748-1317

Boccia, O. ; Chella, F. & Zazzini, P. (2011). Numerical analysis on daylight transmission and thermal comfort in the environments containing devices called "Double Light Pipes", *World Renewable Energy Congress 2011*, Linkoping, Sweden, May 08-13, 2011

Boccia, O. ; Chella, F. & Zazzini, P. (2011). Ventilated Illuminating Wall (VIW): Natural ventilation and daylight experimental analysis on a 1:1 prototype scale model, *World Renewable Energy Congress 2011*, Linkoping, Sweden, May 08-13, 2011

Boccia, O. ; Chella, F. & Zazzini, P. (2011). Ventilated Illuminating Wall (VIW): Natural ventilation numerical analysis and comparison with experimental results, *World Renewable Energy Congress 2011* - Sweden, Linkoping, Sweden, May 08-13, 2011

Carter, D. J. (2002). *The* measured and predicted performances of passive solar light pipe systems, *Light Research Technology*, Vol. 34(1) , pp. 39–52.

Chella, F.; Gentile, E. & Zazzini, P. (2007). Natural light in new underground areas of a historical building: an example of application of double light pipes in preservation of the architectonic heritage, 6th International Conference on Sustainable Energy Technologies SET 2007, pp. 232-237, Santiago de Chile, Chile, September 5-7 2007

Chella, F.; Zazzini, P. & Carta, G. (2006). *Compared numerical and reduced scale experimental analysis on light pipes performances*, 5th International Conference on Sustainable Energy Technologies SET 2006, pp. 263-268, Vicenza, Italy, August 30-September 01, 2006

Jenkins, D. & Muneer, T. (2004). Light-pipe prediction methods, *Applied Energy*, Vol. 79, pp. 77-86.

Jenkins, D., Zhang X. & Muneer, T. (2004). Formulation of semi-empirical models for predicting the illuminance of light pipes, *Energy Conversion & Management*, Vol. 46, pp. 2288-2300.

Jenkins,D. & Muneer, T. (2003). Modelling light-pipe performances-a natural daylighting solution, *Building and Environment*, Vol. 38 pp. 965-972.

Zazzini, P. ; Chella, F.; Scarduzio, A. & (2006). Numerical and experimental analysis of light pipes' performances: comparison of the obtained results, 23th Conference on Passive and Low Energy Architecture, Vol. 2, pp. 219-224, Geneva, Switzerland, September 6-8 2006

Zhang, X. & Muneer, T. (2002). A design guide for performance assessment of solar light pipe, *Light Research Technology*, Vol. 34(2), pp. 149–169.

3

An Approach to Overhang Design, Istanbul Example

Nilgün Sultan Yüceer[*]
Çukurova University,
Faculty of Engineering and Architecture,
Balcali Adana,
Turkey

1. Introduction

Control of solar radiation, by passive solar tools is an important part of building design. External shading device, which is the part of passive solar systems, is an artificial environmental variable or element to control interior solar radiation on the base of desirable orientation of window. Solar heat gain, particularly via fenestration, typically dominates cooling performance (Olgyay, 1957).

The proper application of energy efficient shading devices in new buildings have the technical potential to save 50-70% of total perimeter zone energy use. Therefore, even if only 25-50% of potential could be captured, the economical benefit due to decrease in the size of HVAC plant and the energy consumption make them competitive, with large spin-off benefit on the visual comfort (Data, 2001). Moreover, overshading of the windows reduces daylighting, which results in increasing energy use for artificial lighting or internal heat gains (Littlefair, 1998). In that case, this can be possible with determination of optimum dimension and shape in the shading device design. However, evaluation of many needed variables, such as dimensions of window, solar geometry and climate data in design of shading devices needs a period of long time and include complex process (Szokolay, 1980). This situation makes the designer to pay insufficient attention to the external shading device applications (Ralegaonkar, 2005). In addition, like eaves overhangs, terrace and beams horizontal building elements which are not designed as an external shading device but their shape on the face as an extension of functional or structural devices, shade to transparent surfaces and this shading can negatively influence the thermal performance of the building (Yezioro, 2009). For this aspect, external shading device's shape and dimension should be evaluated carefully at design of the building especially in the Mediterranean climate zone in which solar heat gain is effective.

In shading device design methods, the trigonometric connection between angle of altitude and azimuth of the sun with dimension of window and shading device are taken basic criteria. However, design criteria of shading device is classified in four basic groups as given below (Olgyay, 1957).

---

[*] Corresponding Author

A. Solar geometry data: The formulations and ground plane angle of the sun's yearly motion.

B. Shape and dimension alternatives of window and shading device: The formulations of determining the geometry of window and shading devices.

C. Geographical location and climate data: Climate data and the required comfort data obtained based on the climate data.

D. Function and usage: The shading device's material, detail and usage.

As mentioned above, the most important point in the shading device design is to show the dependence of shading device performance on the base of sun's one-year motion. Thereby, calculating the shading device design parameters with traditional methods requires comparison of various drawings and equations on the base of sun's one-year motion. Computerized design with simulation programming, in offering various numerical and graphical alternatives, is help to the designer by shortening the design period. In this article, computerized simulation program of solar tool is used to analyze shading devices in dimension and shape (Marsh, 2003).

2. Shading device design

Although the computer technology offers infinite numbers of graphics, shading device type and dimension alternatives, and the selection of the optimum solution belong to the designer (Siret, 2004). In other words, performance parameters such as comfort levels, energy saving etc., are those that the decision maker uses to judge the appropriateness of the product (Yezioro,2009). So, the most important part of the shading device design is the selection of the alternatives that provide optimum type and dimension.

This paper describes an approach to simplify and clarify external shading device design. In this regard, Table 1 above, on shading device design criteria has been generated in order to determine how to select the interior comfort shading device dimensions at any site. With the aim of clarifying and facilitating design, "shading device design criteria" have been divided in two main part as seen in Table 1. Some of these criteria vary and some are fixed. In this aspect, the shading mask, which monitors the shading device's performance, and solar radiation, day lighting, climate and comfort graphics, that are used to determine thermal and visual comfort were evaluated as a design criteria to determine the geometry of shading device's optimize dimension and its shape. Then, as seen in the Table 1, by taking into account the criteria interactions with each other, design criteria are selected on the basis of priorities of the climate data of the building's district.

When the design criteria stated in Table 1 are applied to a specific site , parameters related to the dimensions of the window and the shading device could become definite on the direction providing the interior thermal and visual comfort. However, in the building design, dimensions of the windows and shading devices, are used to determine the thermal comfort and sun lighting quality, which sometimes may interact with each other, such as, a shading device that can, on one hand, provide desired shading during the year, and, on the other hand, can reduce day lighting factor (DF) or prevent ventilation (Khaled,2007).

SHADING DEVICE DESIGN CRITERIA			
VARIABLE PARAMETERS		**FIXED PARAMETERS**	
Solar Geometry	Fenestration	Location	Climate
*Angle of the sun VSA: Vertical shadow angle HSA: Horizontal shadow angle ALT: Altitude of the sun ORI: Orientation	*Window dimension *Rear wall dimension *Shading device dimension	*Latitude *Longitude *Time zone *Altitude	*Annual average temperature *Annual wind direction *Annual solar raddiation
	DESIGN	**TOOLS**	
Sun Path Diagrams	Design Options	Comfort Charts & Standards	
*Stereographic *Orthographic *Equidistant	*Material, cost *Function, montaj *Aesthetics, color *Economy	*Bioclimatic chart *Psichrometric chart *ASHREA (Standard 142, 199) *ISO (Standard 7730) *SC% (shading coefficient, ASHREA, DOE)	
MANUAL OR COMPUTER AIDED DESIGN PROCESS			
OPTIMUM DIMENSION-INTERIOR COMFORT-ENERGY EFFICIENCY			

Table 1. Shading device criteria

Furthermore, a shading device dimension, which provides shading in summer months, can provide the same shading in winter months but reduces interior thermal comfort (Szokolay, 1980)). Also, design options of the window and shading devices criteria tools like material, usage, economy and application details may affect the shading devices dimension, too (Miguel, 2008). The colours of the building materials show different behaviour in the aspect of reflecting and absorbing the sun light (Olgyay, 1957). Shading devices, which are made from a material that absorb sun light, used for shading in summer months, may increase heat in the interior space. Besides, window glasses with coatings aimed at reducing incoming solar irradiation, partially reflect or absorb incident solar radiation (Miguel, 2008), so, glazing system effectiveness in controlling solar penetration also affect the shading device in form and material (Alibaba, 2004). In order to resolve the above mentioned problems, shading device dimension and type should be determined first. In this situation, determining optimum device dimensions in a building, that will provide energy efficiency during the year demands series of drawings, which are repeated for every window's dimension, specifying the orientation and geographical position and using the climate data , is essential. Finally, evaluation of all the design criteria stated in Table 1, introduces the design procedure which takes long time and contains complex process. More importantly, it could not always be possible to evaluate all criteria that are in continuous interaction. In this aspect, to give priority to variables or eliminate some of them will be an approach to shading device design. For example, in hot climate zone, interior heat comfort, which can be acquired naturally, could be a priority design criteria for shading device. Furthermore, in cold climate zones, shading device may not be required. Otherwise, protection from sun and heat plays an important role in the Mediterranean zone with a hot climate during the summer, while the problems of areas with cold climate are quite different (Cardinale, 2003). In the residential building district, (which is determined with this procedure)?, selection of one of the effective shading device types and elimination of the others are a solution for making the design easier. Examining the subject critically, only the material, colour, cost and economy that depend on these, can be applied to a shading device with undetermined dimension and shape. From this aspect, in this study, for a selected building area, the necessity of shading device type determination and dimension analyses were made. So, the sun control elements that the users or designers can apply to buildings at Istanbul city, Mediterranean climate zone with latitude 41° North, are seen can not provide the expected energy efficiency because the types and dimensions of external shading devices are not suitable. In this case, priorities of shading device criteria presented in Table 1 are determined on the base of sequence stated below.

2.1 Reasons of the shading device priority that will be applied to the site layout

Location data and climate in Istanbul city, the area of the research, are given in Figure 1. In general, Mediterranean climate conditions are manifested in Istanbul city (Figure 1). In the Mediterranean climate, prevention of solar heat gain is a preferential criterion in building design (Cardinale, 2003). In Istanbul city, in the planned residential areas which were designed without taking into consideration of solar heat gain, the compulsory mechanical cooling systems to obtain interior comfort in summer results in significant or considerate energy consumption.

As stated in Figure 1, the temperatures in June, July, August and September are at the level that affect interior comfort negatively on the base of ASHREA temperature standards (Table 1.). In addition, solar heat gain that are earned in November, December, January, and February is necessarily for interior comfort. In this case, the 6th, 7th, 8th, and 9th months, bright sunshine duration are priority for shading device criteria for latitude 41°N. In this respect, shading blocks which shading device has scanned through in one year period, must stay in June, July, August and September months and between 8:00-16:00 o'clock that the time interval solar heat gain is at the highest level. In Figure1 (yellow line), Latitude 41, are seen the time, month, and orientation that are needed for shading.

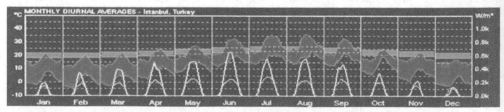

Fig. 1. İstanbul climate (Green: Comfort zone, Yellow : Solar radiation, Red: Average temperature)

2.2 Determination of the shading device type that provides shading only at the required time during one year period

Stereographic diagram in Figure 2 shows shading blocks which includes time, month, and orientation of 180°. In Figure 2, the time, at which the Sun returns from North to South and South to West are considered the base for Istanbul city, which is located at latitude 41°North and longitude 28°West.

The scanned shading blocks in Fig.2 define the required shading block in one year period at latitude 41°.The scanned shading blocks in Figure 2 and Figure 3 can be defined as the "horizontal" type shading device. In this case, horizontal type and south facade (180°) are a priority-shading device for latitude 41.

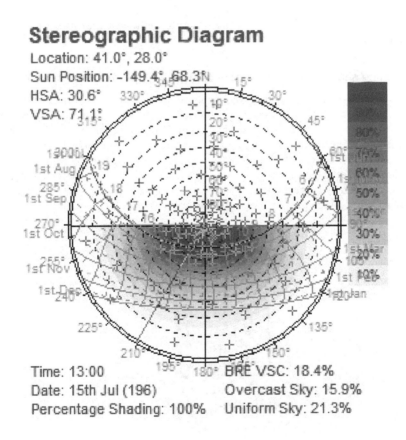

Fig. 2. The scanned shading blocks on diagram (Pink plus: Daylight factor, DF, Blue tone: Shading SC%, Black: Altitude of the sun, ALT, yellow: Date, time)

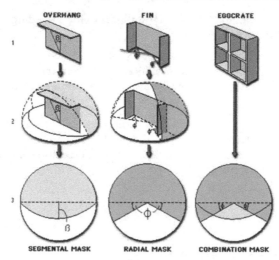

Fig. 3. Shading device types

2.3 Determining shading behaviours of established type dimension alternatives

investigate established The horizontal type shading devices with their dimension behaviours, window, wall and shading device's dimension alternatives, which affect directly performance of shading device, have been determined. In this situation, horizontal device priority-dimension alternatives, like "overhang" priority-dimension alternatives, are applied separately to a fixed "W" type window with width 1.00m., height 1.20m., and the shading blocks are analyzed as seen in Figure 4. Here, overhang is 65 cm wide. Effective Shading Coefficients are seen Figure 5. So average shading coefficients are 15% in winter and 83% in summer which provides energy efficiency.

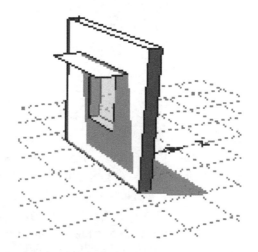

Fig. 4. Overhang

Latitude: 41.0°
Longitude: 28.0°
Timezone: 30.0° [+2.0hrs]
Orientation: 180.0°

Month	Avg.SC	Max.SC	Min.SC
January	13.8%	22.0%	0.0%
February	23.3%	39.0%	0.0%
March	38.2%	64.0%	0.0%
April	69.5%	100.0%	23.0%
May	83.5%	100.0%	48.0%
June	87.3%	100.0%	50.0%
July	78.6%	100.0%	38.0%
August	62.3%	91.0%	23.0%
September	35.5%	58.0%	0.0%
October	21.1%	35.0%	0.0%
November	12.5%	22.0%	0.0%
December	10.4%	20.0%	0.0%
Winter	15.8%	27.0%	0.0%
Summer	83.1%	100.0%	45.3%
Annual	44.7%	62.6%	15.2%

Fig. 5. Effective Shading Coefficients

3. Conclusion

In the design of shading devices, evaluation of many variables data makes the design a complex process. This paper describes an approach to simplify and clarify external shading device design. So a table is generated to determine how to select the criteria for the interior comfort in any building area. Phases and priorities of shading device criteria are analyzed and presented in table 1. Therefore, the design strategies that were followed in this study could be applicable to any building area.

The criteria which are determined from the results of analyses, should be used as a substructure for commercially supplied shading devices like awnings, louvers, shutters and Venetian blinds by designers and users. Furthermore, "the design strategies" which are established in this study are applicable to any building in residential district.

4. References

Data, G.; (2001). *Effect of fixed horizontal louver shading devices on thermal performance of building by TRNSYS simulation*, Renewable Energy, 23, pp.497-507.

Khaled, A. & Al-Sallal; (2007). *Testing glare in universal space design studios in Al-Ain, UAE desert climate and proposed improvements*, Renewable Energy, 32, 6, pp 1033-1044. Yezioro, A.; (2009). *A knowledge based CAAD system for passive solar architecture*, Renewable Energy 34, pp769–779

Littlefair, P.; (1998). *Passive solar urban design: ensuring the penetration of solar energy into the city*, Renewable and Sustainable Energy Reviews, 2, pp.303-326.

Marsh A., J.; (2003)."Solar Tool", Cardiff University. U.K.

Miguel, a. f.; (2008). *Constructal design of solar energy-based systems for buildings*, Energy and Buildings, 40; 6, pp. 1020-1030.Cardinal, N. & Micucci M, Ruggiero F.; (2003).

Analysis of energy saving using natural ventilation in a traditional Italian building, Energy and Buildings, 2003, 35, pp.153-159.Alibaba, H., Z. & Özdeniz, M., B. (2004). *Building elements section system for architects,* Building and Environment, 39, pp307-316.

Olgyay, V.; (1957). *Solar control and shading devices,* Princeton University Press, Princeton NJ, pp. 38-45.

Ralegaonkar R.V., Gupta R., Design development of static sunshade using small scale modelling technique. Renewable Energy, 2005, 30, 6, pp.867-880.

Siret, D. & Houpert, S.; (2004). *A geometrical framework for solving sunlighting problems within CAD systems.* Energy and Buildings, 36, pp.343-351.

Szokolay, S.,V.; (1980). *World solar architecture,* John Wiley and Sons Inc., New York, pp. 7-10.

Yezioro, A.; (2009). *A knowledge based CAAD system for passive solar architecture,* Renewable Energy 34, pp769-779

Section 2

Electricity Application

4

Potential Applications for Solar Photocatalysis: From Environmental Remediation to Energy Conversion

Antonio Eduardo Hora Machado[1,*], Lidiaine Maria dos Santos[1],
Karen Araújo Borges[1], Paulo dos Santos Batista[2],
Vinicius Alexandre Borges de Paiva[1], Paulo Souza Müller Jr.[1],
Danielle Fernanda de Melo Oliveira[1] and Marcela Dias França[1]
[1]*Universidade Federal de Uberlândia, Instituto de Química,
Laboratório de Fotoquímica, Uberlândia, Minas Gerais,*
[2]*Universidade Federal de Goiás, Campus Catalão, Departamento de Química, Catalão, Goiás,*
Brazil

1. Introduction

Taking into consideration the impacts of increasing concern as a result of human activity on the environment in recent decades, different approaches have been developed and proposed to minimize the persistence of organic and inorganic pollutants, not only the dangerous or biorecalcitrant. Industrial waste discharges, those from domestic sewage, and so-called emerging contaminants (pesticides, hormones and drugs), among others, have caused numerous problems for the sustainability of ecosystems (Amat et al.,2011).

In general, environmental problems are largely associated with the disposal of waste into sewers, rivers and eventually into the ocean. The damage caused to biota by these discharges is incalculable (Hermann & Guillard, 2002; Corcoran et al., 2010).

New and effective forms of wastewater treatment are essential to enable a responsible economic development of the planet ensuring its sustainability for future generations (Amat et al.,2011). These processes need to be environmentally safe, providing the elimination of contaminants and not just promoting a phase transfer, ensuring the reuse of water (Hermann & Guillard, 2002;Machado et al., 2003a; Sattler et al.,, 2004; Wojnárovits et al., 2007).

Besides the application in order to minimize the environmental impacts of human action, via photocatalytic processes, semiconductor oxides have also been employed in producing chemical raw materials through specific chemical reactions (Kanai et al., 2001; Murata et al., 2003; Amano et al., 2006; Denmark & Venkatraman, 2006; Hakki et al.,2009; Swaminathan & Selvam, 2011; Swaminathan & Krishnakumar, 2011), in the conversion of solar energy into electricity (Prashant, 2007; Patrocínio et al., 2010; Huang et al., 2011; Zhou et al., 2011) and

---

* During his leave as Visiting Professor in the Departamento de Química at the Campus Catalão of Universidade Federal de Goiás.

production of hydrogen for subsequent generation of energy (Jing et al., 2010; Kim & Choi, 2010; Melo & Silva, 2011).

2. Advanced Oxidation Processes (AOP)

Advanced Oxidation Processes (or Advanced Oxidative Technologies) stand out among the new technologies potentially useful for the minimization of environmental impacts to biota (Ismail et al., 2009), and, among these technologies, are the photocatalytic degradation of contaminants in the environment, especially using solar radiation (Martin et al., 1995; Ziolli & Jardim, 1998; Machado et al., 2003a; Duarte et al.,2005; Augugliaro et al., 2007; Machado et al., 2008). They are characterized by being able to degrade a wide range of organic contaminants into carbon dioxide, water and inorganic anions through reactions involving oxidizing species, particularly hydroxyl radicals which have a high oxidizing power (E^o=2.8 V) (Nogueira & Jardim, 1998; Machado et al., 2003a; Machado et al., 2008; Kumar & Devi, 2011).

Among the AOP can be cited processes involving the use of ozone, hydrogen peroxide, catalytic decomposition of hydrogen peroxide in acid medium (Fenton or/and photo-Fenton reactions), and semiconductors such as titanium dioxide (heterogeneous photocatalysis) (Nogueira & Jardim, 1998; Kumar & Devi, 2011). The heterogeneous photocatalysis is considered one of the most promising advanced oxidation technologies. In heterogeneous photocatalytic processes, highly oxidizing reactive oxygen species (ie hydroxyl radicals, superoxide radical-ions, etc.) are generated from interaction between the semiconductor electronically excited, oxygenated species and other substrates (Andreozzi et al., 1999; Fujishima et al., 2007; Machado et al., 2008; Kumar & Devi, 2011).

The solar photocatalysis deserves special attention, since the sun is a virtually inexhaustible source of energy at no cost (Machado et al., 2008; Amat et al., 2011).

2.1 Heterogeneous photocatalysis

The great potential of heterogeneous photocatalysis has been demonstrated mainly in the treatment of industrial effluents and wastewater through the degradation of contaminants (Malato et al., 1997; Andreozzi et al., 1999; Malato et al., 2002; Sattler et al., 2004a, 2004b; Duarte et al., 2005; Pons et al., 2007; Palmisano et al., 2007a; Machado et al., 2008). A significant number of these studies have focused on the photocatalytic properties of TiO_2, suggesting a promising use of this material in heterogeneous photocatalysis (Mills & Hunte, 1997; Malato et al., 2002; Mills et al., 2002; Machado et al., 2003a; Machado et al., 2003b; Sattler et al., 2004a, 2004b; Duarte et al., 2005; Palmisano et al., 2007a; Pons et al.,2007; Machado et al., 2008; Oliveira et al., 2012).

The potential of heterogeneous photocatalysis has been demonstrated in studies originally reported by Fujishima and Honda (Fujishima & Honda, 1971, 1972). The photoactivation of a semiconductor is based on its electronic excitation by photons with energy greater than the band gap energy. This tends to generate vacancies in the valence band – VB (holes, h+) and regions with high electron density (e-) in the conduction band – CB (Hoffmann et al., 1995; Nogueira & Jardim, 1998; Kumar & Devi, 2011). These holes have pH dependent and strongly positive electrochemical potentials, in the range between +2.0 and +3.5 V, measured against a saturated calomel electrode (Khataee et al., 2011). This potential is sufficiently positive to generate hydroxyl radicals (**HO·**) from water molecules adsorbed on the surface of the semiconductor (eqs. 1-3). The photocatalytic efficiency depends on the competition between the formation of pairs of electrons and holes in semiconductor surface and the recombination of these pairs (eq.4) (Nogueira & Jardim, 1998; Ziolli & Jardim, 1998; Ni et al., 2011).

$$TiO_2 + hv \rightarrow TiO_2 \, (e^-_{CB} + h^+_{VB}) \tag{1}$$

$$h^+ + H_2O_{ads.} \rightarrow HO\cdot + H^+ \tag{2}$$

$$h^+ + OH^-_{ads.} \rightarrow HO\cdot \tag{3}$$

$$TiO_2 \, (e^-_{CB} + h^+_{VB}) \rightarrow TiO_2 + \Delta \tag{4}$$

The electrons transferred to the conduction band are responsible for reducing reactions, such as the formation of gaseous hydrogen and the generation of other important oxidizing species such as superoxide anion radical. In the case of TiO_2, the band gap energy, E_g, is between 3.00 and 3.20 eV (Hoffmann et al., 1995; Palmisano et al., 2007a; Jin et al., 2010; Kumar & Devi, 2011). This process can be viewed schematically in **Fig. 1.**

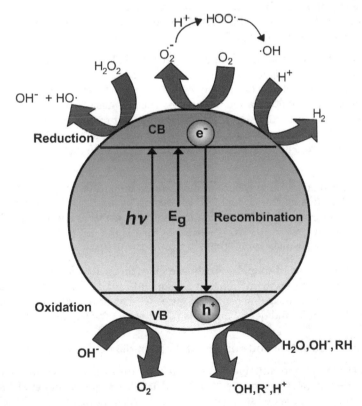

Fig. 1. General scheme for some primary processes that occur after photoactivation of a semiconductor and for photocatalytical production of gaseous hydrogen by decomposition of water.

The production of reactive species by a TiO_2 photocatalyst is influenced by a series of factors, such as surface acidity and pH of the reaction medium, control of the kinetic of recombination of charge carriers, interfacial electron-transfer rate, optical absorption of the semicondutor, phase distribution, morphology, specific surface area and porosity (Hoffmann et al., 1995; Furube et al, 2001; Diebold, 2003; Carp et al., 2004; Kumar & Devi (2011).

The reactions (1) to (4) combined with other (Hoffmann et al., 1995; Machado et al., 2008; Kumar & Devi, 2011) give an approximate view of the chain reactions that compose a heterogeneous photocatalytic process.

Different semiconductors are able to trigger the heterogeneous photocatalytic processes. Other in addition to TiO_2 are: CdS, ZnO, ZnS, and Fe_2O_3 (Nogueira & Jardim, 1998).

TiO_2 stands in front of others for its abundance, low toxicity, good chemical stability over a wide pH range, photosensitivity, photostability, insolubility in water, low cost, chemical inertness, biological and chemical inertness, and stability to corrosion and photocorrosion (Martin et al., 1995; Augugliaro et al., 2007). However, its band gap energy limits, in principle, its application in photocatalytic processes induced by solar radiation, since the radiation incident on the biosphere consists of approximately 5 % UV, 43 % visible and 52 %, harvesting infrared (Kumar & Devi, 2011).

The introduction of changes in the crystalline structure of TiO_2 through the introduction of dopant ions and/or modifying ions and associations between TiO_2 and other semiconductor oxidesin order to expand the use of incident radiation, is particularly important if the aim is to use solar radiation in photocatalytic processes. The synthesis of new materials based on TiO_2 has resulted in substantial progress towards the improvement of the photocatalytic activity of this semiconductor (Imhof & Pine, 1997; Cavalheiro et al.,2008; Eguchi et al., 2001; Agostiano et al., 2004; Machado et al., 2008; Zaleska et al., 2010; Batista, 2010; Machado et al., 2011b).

Titanium dioxide can be found in nature in the form of three different polymorphs: Anatase, Rutile and Brookite (Hanaor & Sorrell, 2011; Khataee et al., 2011; Kumar & Devi, 2011). Among these polymorphs, the thermodynamically more stable is the rutile, which can be obtained from the conversion of anatase, which in turn is the most photoactive polymorph (Hoffmann et al., 1995; Khataee et al., 2011).

Technological applications of titanium oxide are quite large. In addition to the previously described, TiO_2 has been used in filters to absorb ultraviolet radiation (sunscreens, for example), pigments, in chemical sensors for gases (Pichat et al., 2000), as constituents of ceramic materials for bone and dental implants (Chen et al., 2008), among others.

2.2 Changes in the structure and surface of titanium dioxide

Strong light absorption and suitable redox potential are prerequisites for photocatalytic reactions. Growing interest has focused on doped TiO_2 catalysts (Ohno et al., 2003; Luo et al., 2004; Li et al., 2005; Labat et al., 2008; Yang et al., 2008; Long et al., 2009; Zhang et al., 2010; Zaleska et al., 2010; Iwaszuk & Nolan, 2011; Long & English, 2011; Spadavecchia et al., 2011; Kumar & Devi, 2011;), however current achievements are still far from the ideal goal.

In order to extend the photocatalytic activity in the region of visible light, and in order to achieve a better use of solar radiation, several approaches have been proposed for tuning the band gap response of titania to the visible region. Doping or incorporate trace impurities in the structure of TiO_2 in order to obtain materials with photocatalytic activity maximized in the visible region are strategies widely used (Ohno et al., 2003; Li et al., 2005; Zaleska et al., 2010). These strategies include doping with transition metals (Nogueira & Jardim, 1998; Yamashita et al., 2001; Cavalheiro et al., 2008; Zaleska et al., 2010), nonmetals (Ohno et al., 2003; Li et al., 2005), and the inclusion of low-valence ions on the surface of the semiconductor (for example, Ag^+, Ni^{3+}, V^{3+} e Sc^{3+}). Certain metals, when incorporated to titanium dioxide, are able to decrease the band gap, making possible in some cases its application in solar photocatalysis. Furthermore, they can contribute to minimize the electron-hole recombination, increasing the photocatalytic efficiency of the semiconductor (Zaleska et al., 2010).

Coupling of two photocatalysts has also been considered effective for improvement of photocatalytic efficiency. As example, nitrogen doped TiO_2 coupled with WO_3 and after loaded with noble metal, resulted in a material with improved photocatalytic efficiency (Yang et al., 2006).

2.2.1 Synthesis of TiO_2

We have performed the synthesis of titanium dioxide using different methodologies (Batista, 2010; Oliveira, 2011). A modification was introduced in the methodology of the synthesis by precipitation of TiO_2 using titanium tetraisopropoxide as precursor suggested by Batista (Batista, 2010). It consists in making the whole process, since the solubilization of the precursor in 2-propanol, always under the action of ultrasound. The solid obtained was dried at 60 ° C and subjected to heat treatment at 400 ° C. This new photocatalyst has been adopted in our most recent studies since it has shown impressive photocatalytic activity in the mineralization of different organic substrates (Machado et al., 2011b). As a result, we have studied the introduction of modifications in order to enlarge it, especially expanding it to the visible.

After annealing, the semiconductor was highly crystalline, being only anatase with average crystallite size around 12 nm, estimated from the line width obtained for the peak of greatest intensity in XRD (**Fig.2**). For a semiconductor synthesized according to a similar methodology adopted by Batista, the minimum crystallite size obtained was equal to 22 nm (Batista, 2010).

From the curves of diffuse reflectance, the band gap of the synthesized TiO_2 and TiO_2 P25 Degussa were estimated. For this, we used Tauc´s method (Wood & Tauc, 1972). For the synthesized TiO_2 was obtained a value equal to 3.18 eV while for TiO_2 P25 Degussa the estimated band gap was equal to 3.20 eV, in agreement with the value described by many authors (Hoffmann et al., 1995; Machado et al., 2008; Batista, 2010). The earlier versions obtained by precipitation, reported by Batista in his DSc Thesis (Batista, 2010) showed no photocatalytic activity due to its proper degree of aggregation and in some cases limited surface area. Most likely, due to the significant aggregation observed in semiconductor synthesized by Batista (2010), the recombination of charge carriers was more favored at the expense of photocatalytic reactions. It is very likely that the introduction of ultrasound in the synthesis process resulted in significant increase in the dispersion of the particles formed during the formation of critical nuclei, resulting in the precipitation of particles with minimal or no aggregation. Morphological characterization of this new photocatalyst is ongoing.

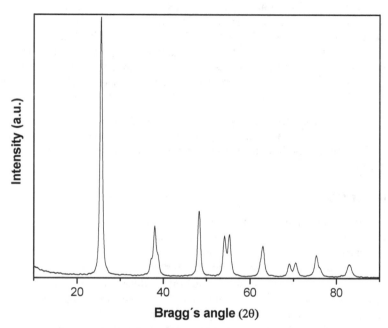

Fig. 2. X-ray diffraction patterns of TiO$_2$ synthesized by solubilization of titanium tetraisopropoxide in 2-propanol and subsequent hydrolysis and precipitation by slow addition of ultrapure water.

The mineralization of food dye trartrazine, C.I. 19140, mediated by this new photocatalyst is presented as an example. It was promoted at pH 3, using 100 mg/L of photocatalyst, in experiments on laboratory scale, using as radiation source a 400 W high pressure mercury vapor lamp. 4 L of the model effluent were used per experiment. Hydrogen peroxide (166 mg/L) was added as an extra font of radicals (Machado et al., 2003a). The results were compared to the obtained under the same conditions using TiO$_2$ P25 Degussa as photocatalyst. Additionally, all photolysed samples underwent the following tests: pH monitoring, spectrophotometric measurements through the use of a UV/VIS dual beam Shimadzu UV-1650PC spectrophotometer. The aliquots collected in the experiments in the presence of the photocatalyst were filtered using Millipore filters (0.45 μm of mean pore size) to remove suspended TiO$_2$ before the measurements. The experimental setup is similar to that described in previous studies (Machado, 2003; Oliveira, 2012).

After 120 minutes of reaction, 52% of mineralization was reached with the use of the synthesized TiO$_2$. For TiO$_2$ P25-mediated degradation, the mineralization was 84% under the same conditions In the absence of H_2O_2, the levels of mineralization were respectively 24 and 38% for the synthesized TiO_2 and TiO_2 P25 **(Fig. 3)**. The mineralization was estimated from measurements of dissolved organic carbon using a Shimadzu TOC-VCPH Total Organic Carbon Analyzer.

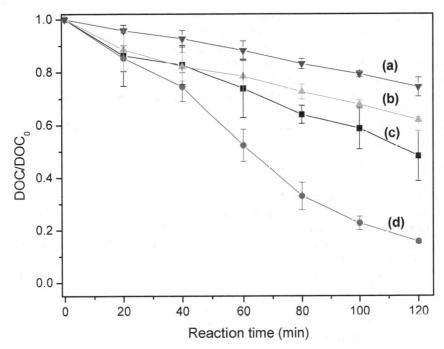

Fig. 3. Mineralization of tartrazine by heterogeneous photocatalysis using: (a) TiO_2 synthesized in reaction in the absence of H_2O_2; (b) TiO_2 P25 in reaction in the absence of H_2O_2; (c) TiO_2 synthesized, in reaction in the presence of H_2O_2, and (d) TiO_2 P25 reaction in the presence of H_2O_2.

The changes introduced during the solubilization and synthesis process itself should have been enough to guarantee a level of ordering of the particles formed. The final product after thermal treatment of the oxide formed proved to be 100% anatase.

2.2.2 Photocatalysts based on the association between a photosensitizing dye and a semiconductor oxide

Electron transfer at the interface between a photoactive species and the semiconductor surface is a fundamental aspect for organic semiconductor devices (Grätzel, 2001; Ino et al., 2005). Certain photoactive compounds has proven to be able, when electronically excited, to inject electrons in the conduction band of semiconductors (Grätzel, 2001; Ino et al., 2005; Rehm et al., 1996; Nazeeruddin et al., 1993; Asbury et al., 2001; Krüger et al., 2001; Argazzi et al., 1998; Xargas et al., 2000; Tennakone et al., 1997; Sharma et al., 1991; Hao et al., 1998; Chen et al., 1997; Wu et al., 2000), increasing the performance of dye-sensitized solar cells. In particular, ultrafast charge separation led by electron injection from electronically excited photoactive molecules to the conduction band of a wide-gap metal oxide, and a good electronic coupling between dye molecules and surface of the substrate are key steps for improving the performance of these materials (Rehm et al., 1996; Nazeeruddin et al., 1993; Asbury et al., 2001). In the dye sensitization process, dye gets excited rather than the TiO_2 particles to appropriate singlet and triplet states, being subsequently converted to cationic dye radicals after electron injection to TiO_2 CB (Benko et al., 2002). The electrons injected to TiO_2 CB react with the preadsorbed O_2 to form oxidizing species (superoxide, hydroperoxyl and hydroxyl radicals) which combined to the species produced from photoexcited TiO_2, induce oxidative reactions (Wu et al., 1998). Thus TiO_2 plays an important role in electron-transfer mediation, even though TiO_2 itself is not excited. A photodegradation mechanism of dyes under visible irradiation without TiO_2 photoexcitation was recently presented by Kumar & Devi (2011). The formation of singlet oxygen has been reported in some cases (Stylidi et al., 2004).

The association between photosensitizing dyes and oxides semiconductors with photocatalytic activity constitutes a strategy for obtaining more efficient photocatalysts for a wider range of applications. These photosensitizing dyes, when excited by photons of lower energy, allow the injection of electrons from these species to the conduction band of the semiconductor increasing the concentration of charge carriers (Benkö et al., 2002; Sharma et al., 2006; Machado et al., 2008; Shang et al., 2011; Kumar & Devi, 2011). The electrons, in turn, can be transferred to reduce organic acceptors adsorbed on the catalyst surface (Machado et al.,2008). Thus, the photocatalyst composites containing a photosensitizing dye associated with the photoactive semiconductor have, in general, improved photocatalytic activity. The possibility of utilization of solar radiation, because they have the range of absorption expanded to the visible, makes it possible achieve important contributions in solving problems concerning effluent treatment (Machado et al.,2008). Machado and coworkers (2003b; 2008; 2011; Duarte et al., 2005) have studied composites prepared by the association between zinc phthalocyanine (ZnPc) and titanium dioxide, obtained by coating TiO_2 particles using a solution of zinc phthalocyanine followed by controlled drying of the organic suspension (Machado et al., 2008). These materials have been intensively characterized (Machado et al., 2008; Batista et al., 2011). A decrease between 20 and 30% in the specific surface area (SSA) is verified for the composites when compared to the TiO_2 P25 (Machado et al., 2008; Oliveira et al., 2011; Batista et al., 2011). This difference should be as a result of the incorporation of ZnPc aggregates on the surface of the semiconductor. The changes in the specific area caused by the incorporation of zinc phthalocyanine do not imply distortions in the crystal structure (Machado et al., 2008). Scanning tunneling microscopy of different metal phthalocyanines confirm that the above mentioned aggregates are adsorbed onto the semiconductor surface (Qiu et al., 2004).

For these composites, the surface sensitization by electron transfer via physisorbed ZnPc should compensate the decrease in surface area, increasing the efficiency of the photocatalytic process. It should be emphasized that the extended range of wavelengths shifted to the visible region of the electromagnetic spectrum, which is capable of positively influencing the electron transfer between the excited dye and the semiconductor conduction band tends to improve electron–hole separation (Machado et al., 2008; Carp et al., 2004;

Wang et al., 1997; Shourong et al., 1997; Zhang et al., 1997; Zhang et al., 1998). These composites have shown to be better photocatalysts for wastewater decontamination, mainly mediated by visible light, than pure TiO_2 (Machado et al., 2003b; Duarte et al., 2005; Machado et al., 2008; França, 2011; Oliveira et al., 2012), performance that remains even when reused (Machado et al., 2008).

The zero point charge pH (pH$_{ZPC}$) was estimated for TiO_2 P25 and a composite containing 1.6% m/m of ZnPc by zeta potential measurements, carried out in a disperse suspension using a Zetasizer Nano ZS90. The estimated value for the composite, pH$_{ZPC}$ = 5.50, lower than the one for P25 (pH$_{ZPC}$ = 6.25) suggests a differentiated behavior for the composite since its surface is negatively charged in a pH range in which P25 is still with the surface positively charged. The value measured for TiO_2 P25 agrees with the reported in the literature (Hoffmann et al., 1995). The morphological characteristics of both samples were investigated by SEM, carried out in a Philips XL-30 microscope coupled to a field emission gun and a EDX analytical setup. The micrographs show the occurrence of macro-aggregates in the composite and spherical particles around 25 nm in P25. The estimated concentration of ZnPc on P25 surface is around 1.6%, confirmed by EDX measurements (Batista et al., 2011). Also, the thickness of ZnPc coating, homogeneity, and aggregation on the TiO_2 composite surface were evaluated by TEM using a Philips CM-120 microscope. The improvement of visible light absorption in TiO_2/ZnPc and electronic surface properties of this composite (Machado et al., 2008) are responsible for an almost three times faster mineralization of Ponceau 4R (C.I. 16255), an azo dye employed in the food industry, when compared with the result obtained using only TiO_2 P25, and still much higher than the presented by the other TiO_2-based photocatalysts (Oliveira et al., 2012). This dye is classified as a carcinogen in some countries and is currently listed as a banned substance by U.S. Food and Drug Administration (FDA).

The highest photocatalytic activity of TiO_2/ZnPc 1.6% seems to be the result of synergism between the photocatalytic characteristics inherent to TiO_2 P25 with the redox properties and charge transport of ZnPc Frenkel's "J" aggregates on the semiconductor surface (Fidder et al., 1991; Kim et al., 2006; Machado et al., 2008; Machado et al., 2011a). The sensitization of TiO_2 P25, induced by zinc phthalocyanine aggregates was effective in producing more active photocatalysts.

Fig. 4 presents the diffuse reflectance spectra (DRS) of ZnPc, TiO_2 and some of the studied TiO_2/ZnPc composites.

Unlike what occurs with TiO_2 (**Fig. 4a**), for composite materials obtained by the association between TiO_2 and ZnPc there is a significant electronic absorption for wavelengths above 390 nm. Comparison between the graphs presented in As can be seen in **Fig. 4 (a to e)**, the UV-Vis absorption spectrum (DRS) of these composites is not the result of an additive effect between the absorption spectra of the precursors. The absorption spectra of the composites are quite different from the typical absorption profiles of TiO_2 (**Fig. 4a**) and pure ZnPc in the solid state (**Fig. 4f**) or even in very dilute liquid solutions (Miranda et al., 2002).

The absorption spectrum of these composites is characterized by an intense absorption band below 460 nm, and a large, intense and non structured absorption band above 475 nm. Both bands are most probably the result of superposition of electronic states of TiO_2 and ZnPc aggregates.

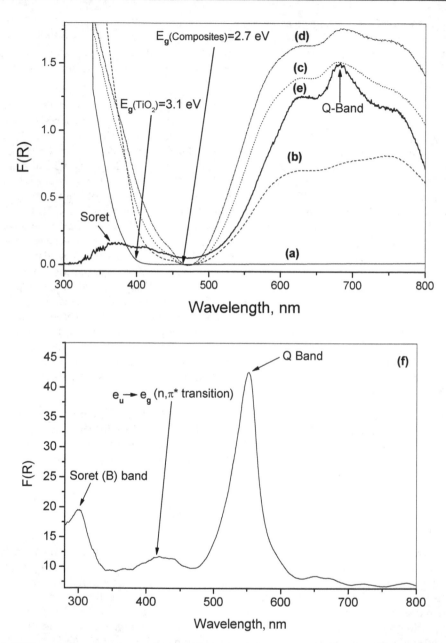

Fig. 4. Diffuse reflectance spectra (DRS) of TiO_2 and $TiO_2/ZnPc$ composites, prepared with different percent in mass of ZnPc. TiO_2 P25 (a) and composite containing: 1.0% of ZnPc (b); 2.5% of ZnPc (c); 5.0% of ZnPc (d); composite containing 2.5% of ZnPc, using TiO_2 P25 as reference (e) and DRS of pure ZnPc (f). Barium sulphate was used as reference for (a) to (d) (Machado et al., 2008).

In **Fig. 4e** the shape of the bands in the ultraviolet and visible portions of the electronic spectrum of the composite containing 2.5% m/m of ZnPc, obtained using TiO_2 as reference, is very different from that observed for pure ZnPc in the solid state, **Fig. 4f**. In the visible, it presents a large and intense three peak band centered by a red shifted Q band, with maximum at 683 nm. The batochromic shift of the absorption maximum associated to Q band, suggests the occurrence of Frenkel's J aggregates of ZnPc (**Fig. 5**) in the composites (Köhler & Schmid, 1996; Eisfeld & Briggs, 2006; Chen, Z. et al., 2008), which agrees with results of a theoretical study employing methods of Density Functional Theory on the formation of aggregates of zinc phthalocyanine (Machado et al., 2011a). The bathochromic shift of the absorption maximum of the Q band highlights the differentiated nature of these compounds against pure TiO_2 and ZnPc. The Soret (B) band also presents a different shape compared to its equivalent in pure ZnPc in the solid state (**Fig. 4f**), and is red shifted. The spectrum of **Fig. 4e** is very similar to the absorption spectrum for a flash-evaporated ZnPc thin film deposited on a glass substrate (Senthilarasu et al., 2003), in which the two energy bands characteristic of phthalocyanines are evident, one in the region between 500 and 900 nm, with an absorption peak at 690 nm, related to the Q band, and the other, very intense, at 330 nm, attributed to Soret (B) band (Meissner & Rostalski, 2001), similar to that reported for the absorption spectrum for thin films of Magnesium Phthalocyanine (Mi et al., 2003). The unstructured band in the visible and the red shifted Q band of these composites can be attributed to the strong intermolecular interactions due to ZnPc aggregation ($ZnPc_{agg}$), resulting in coupling effects of excitons on the allowed transitions, with significant effects on the mobility of charge carriers (Hoffmann, 2000).

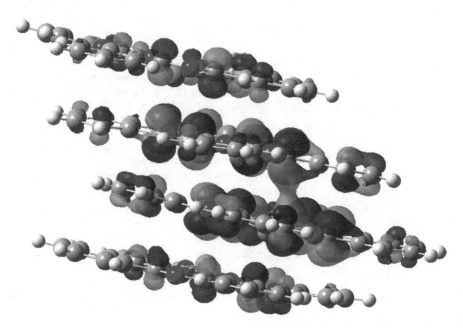

Fig. 5. Representation of the molecular structure of a Frenkel's J aggregate of ZnPc formed by four grouped individual molecules, indicating the sharing the same ligand MO between the ZnPc 2 and 3, in the HOMO (Machado et al., 2011a).

Fig. 4f presents the diffuse reflectance spectrum of pure ZnPc. The intense absorption peak at 552 nm, is related to the Q band and is attributed to very intense $\pi \to \pi^*$ transitions (Leznoff & Lever, 1990). The Soret band presents an absorption maximum at 301 nm. A low intensity and non structured absorption band with the absorption peak centered at 416 nm, is related to an $n \to \pi^*$ transition involving the e_u azanitrogen lone pair orbital with the e_g LUMO (Ricciardi et al., 2001). A set of three very small intensity low energy bands, above the Q band, can also be observed.

The E_g value for the $TiO_2/ZnPc$ composites, 2.7 eV, lower than the estimated for pure TiO_2 (Hoffmann et al., 1995), has a value similar to the estimated for iron (II) phthalocyanine excitons (2.6 eV) in $TiO_2/FePc$ blends (Sharma et al., 2006) and other metal phthalocyanine associated to semiconductor oxides (Iliev et al., 2003). For ZnPc thin films, Senthilarasu et al. assigned an E_g of 1.97 eV (Senthilarasu et al., 2003) with a directly allowed optical transition, near the value estimated for the peak absorption Q-band (2.25 eV) of pure ZnPc in the solid state (**Fig. 4f**). The E_g for the composites might be related to the coupling between TiO_2 and ZnPc electronic states and their positive implications. Similar to $TiO_2/FePc$ blends (Sharma et al., 2006) and ZnPc thin films (Ino et al., 2005; Senthilarasu et al., 2003), the photoexcitation of ZnPc aggregates should result in the formation of $e^-/ZnPc^+$ pairs, followed by electron transfer from ZnPc excitons to the conduction band of bulk TiO_2, which explains at least in part the improved photocatalytic activity observed for some of the $ZnPc/TiO_2$ composites (Machado et al., 2008; Oliveira et al., 2012). Sharma et al. reported charge separation after photo-excitation of $TiO_2/FePc$ composite film due to charge transfer from FePc to TiO_2 resulting in $FePc(h^+)$ and $TiO_2(e^-)$ (Sharma et al., 2006). Additionally, they reported that the charge transport and the current leakage through FePc films and the photo-generation are due to the efficient dissociation of exciton at the donor–acceptor interface of the bulk, and that the higher holes mobility in the organic material layer, combined with lower conductance leakage, leads to the more efficient collection of photo-generated carriers. Thus, the electronic coupling strength between donor and acceptor is one of the critical conditions to ensure the occurrence of such electron transfer (Ino et al., 2005; Rehm et al., 1996; Senthilarasu et al., 2003; Meissner & Rostalski, 2001).

The spectrum presented in **Fig. 4e** is very similar to the absorption spectrum for a flash-evaporated ZnPc thin film deposited on a glass substrate (Senthilarasu et al., 2003), in which the two energy bands characteristic of phthalocyanines are evident, one in the region between 500 and 900 nm, with an absorption peak at 690 nm, related to the Q band, and the other, very intense, at 330 nm, attributed to Soret (B) band (Meissner & Rostalski, 2001), similar to that reported for the absorption spectrum for thin films of Magnesium Phthalocyanine (Mi et al., 2003).

2.3 Solar photocatalysis using a compound parabolic concentrator (CPC) reactor

2.3.1 Design and construction of a CPC reactor

The study of new technologies has now focused on decontamination methods feasible alternatives that are environmentally friendly, and allow its application in large scale, with easy operation and low cost.

The economic use of AOPs based on the use of solar radiation in the treatment of wastewater has been proposed for their low cost, especially in regions with high insolation

(Malato et al., 2002; Machado et al., 2003; Sattler et al., 2004a, 2004b; Machado et al.,2004; Palmisano et al., 2007; Machado et al., 2008; Torres et al., 2008; Li et al., 2009). Literature reports suggest that the reactors most suitable for application in solar photocatalysis are CPC type (Malato et al., 1997; Malato et al., 2002; Sattler et al., 2003a, 2003b; Machado et al.,2004; Duarte et al., 2005; Machado et al.,2008).

CPC reactors are static collectors of solar radiation with reflective surfaces in the form of involute positioned around cylindrical tubes, **Fig. 6**. Reflectors with this geometry allows the pock up of solar radiation, either by direct incidence, as the diffuse radiation, directing it to a glass tube through which circulates the effluent to be treated (Duarte et al., 2005).

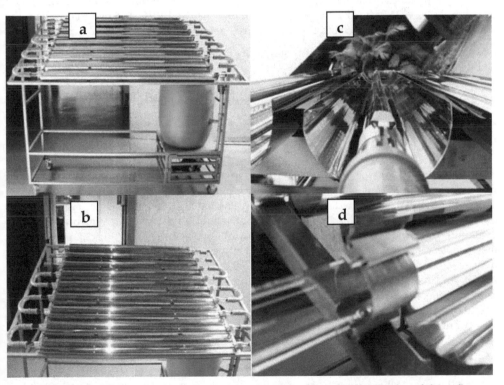

Fig. 6. Representation in two angles (a and b) of a CPC reactor, detailing one of the reflectors in the form of involute (c), and pipes the fixed to the body of the reactor (d).

Our CPC reactor was designed to process up to 150 L of effluent, This reactor consists in a module with an aperture of about 1.62 m², elevation angle adjusted to the latitude of Uberlândia, Brazil (19° S), ensuring a better use of incident radiation. The reflecting surface contains 10 borosilicate glass tubes (external diameter 32 mm, wall thickness of 1.4 mm, and length of 1500 mm), mounted in parallel and connected in series, each on double parabolic shaped inox steel reflector surfaces (Duarte et al., 2005). A centrifugal pump of 0.50 HP with rotor and housing made in inert material has been used to ensure a flow of 2 m³/h.

The flow of effluent in tubular reactors is usually turbulent, which may cause loss of efficiency in the capture of solar radiation. However, this difficulty can be minimized during the design of the reactor, and the use of balanced amounts of the catalyst, in the case of heterogeneous photocatalysis, so as to guarantee a uniform flow and a good dispersion of the photocatalyst in the effluent to be treated, minimizing possible effects of co-absorption of the incident radiation (Duarte et al., 2005). Non-uniform flows implies in non uniform residence times that can lower efficiency compared to the ideal conditions (Koca & Sahin, 2002). In the case of the heterogeneous processes with photocatalyst powder in suspension, sedimentation and depositing of the catalyst along the hydraulic circuit should be avoided and turbulent flow in the reactor needs be guaranteed. Reynolds's number varying between 10 000–50 000 ensures fully turbulent flow and avoids the settlement of the photocatalyst particles in the tubes (Malato et al., 2002). In our project, the Reynolds' number were defined as being $Re_{glass} = 34,855.4$ and $Re_{PVC} = 40,070.0$, for glass and PVC, the materials where the effluent with the photocatalyst in suspension circulate.

Details of the project of a CPC reactor similar to the built in our laboratory are available in Duarte et al., 2005.

2.3.2 Photocatalytic degradation of organic substrates using solar radiation

2.3.2.1 Degradation of organic matter present in a model-effluent simulating the wastewater produced by a pulp and paper industry, using TiO$_2$ P25 and the composite TiO$_2$/ZnPc 2.5% m/m

The performance of the studied composites to degrade organic matter present in wastewaters, in reactions mediated by solar irradiation, and the possibility of reuse of such photocatalysts, was evaluated monitoring the consumption of the organic matter content during the treatment of three 50 L batches of a model effluent (an aqueous solutions containing 160 mg L^{-1} of a sodium salt of lignosulphonic acid (Sigma-Aldrich), possessing a mean molecular mass of 52,000 D. The reactions were done at pH 3, with the addition of hydrogen peroxide (30 mg L^{-1}), used as additional source of reactive species (Machado et al., 2003a), and monitored by chemical oxygen demand (COD) analysis of aliquots of effluent samples collected at different accumulated doses of UV-A radiation (this option was due to operational limitations. However, the spectral pattern of the visible light does not change significantly during the execution of the experiments). To evaluate the observed (global) reaction kinetics, the temporal variations were substituted by the UVA accumulated dose, which warrants the reproducibility of these experiments under different latitude and weather conditions. The incident UV-A radiation was monitored using a Solar Light PMA-2100 radiometer. All reactions were stopped when the accumulated dose of UVA reached 900 kJ m^{-2} (Machado et al., 2008). This corresponds to about 3 hours of sunlight on a sunny day, or 5 to 6 hours during a cloudy day with moderate to high nebulosity in Uberlândia, MG, Brasil (Duarte et al., 2005).

The COD measurements considered the Environmental Protection Agency (EPA) recommended method (Jirka & Carter, 1975).

A same sample of the photocatalyst (100 mg per liter of effluent), containing initially 2.5% of ZnPc, was used to treat the three effluent batches. The treatment of each batch was performed using a CPC (Compound Parabolic Concentrator) reactor (Duarte et al., 2005).

As reference, an additional effluent batch was treated under similar conditions using pure TiO_2 P25 as photocatalyst.

The degradation of the sodium salt of lignosulphonic acid (LSA) suggests higher photocatalytic efficiency for the TiO_2/ZnPc composite. **Fig. 7** shows a more effective LSA degradation under the action of TiO_2/ZnPc, which increases with reuse, with significant changes in the degradation profile due to the use of the recovered composite While under the action of TiO_2 P25 was reached 60% degradation, under the same conditions, with the unused composite, the degradation reached 96%. For the composite in both the first and second reuse, the degradation of the LSA was about 90%. The change in profile suggests that other processes, less likely to occur before, became important for the overall reaction (Machado et al., 2008). The production of singlet oxygen by photosensitization from ^{3}ZnPc*, for example, is an event plausible if the level of aggregation of ZnPc is reduced. The formation of singlet oxygen has been reported in some cases (Stylidi et al., 2004).

On the other hand, the better hydration of the surface of the composite due to the increasing number of cycles of use, should favor reactions from the valence band.

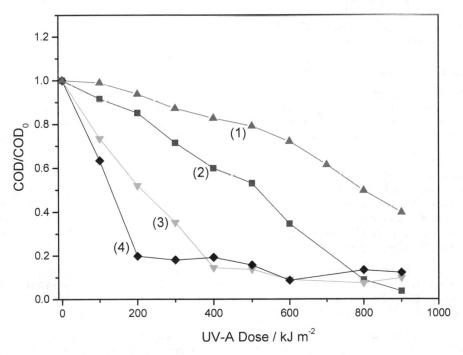

Fig. 7. Degradation of the organic load present in 50 L of a model waste water containing LSA monitored in terms of relative chemical oxygen demand (COD/COD_0), induced by: (1) TiO_2 P25; (2) TiO_2/ZnPc 2.5%; (3) TiO_2/ZnPc 2.5% in the first recycling; (4) TiO_2/ZnPc 2.5% in the second recycling.

Despite the fact that part of ZnPc adsorbed to the surface of TiO_2 P25 may have been degraded during the photocatalytic process, surprisingly, the photocatalytic efficiency of the composite did not decrease when reused. Results suggest that the composite can be reused at least five times before making any significant loss of photocatalytic efficiency.

2.3.2.2 Photocatalytic degradation of paracetamol using solar photocatalysis

Conditions were evaluated to promote the mineralization of paracetamol (or acetaminophen), a known emerging contaminant (Daughton, C. G. & Ternes, T. A., 1999; Bound, J. P. & Voulvoulis, N., 2004; Jones, O. A. H., Voulvoulis, N. et al. , 2007; Nikolaou, A., Meric, S. et al., 2007; Zhang, X. *et al.*, 2008; Bartha, B. *et al.*, 2010), employing heterogeneous photocatalysis mediated by TiO_2/ZnPc 2.5% m/m, under the action of solar radiation. The results were compared with process under similar conditions, using TiO_2 P25 as photocatalyst.

Firstly, to find the best experimental conditions, the influence of hydrogen peroxide concentration and pH was evaluated in the photocatalytic reactions mediated by 100 mg L^{-1} of photocatalyst, on a laboratory scale using an experimental setup already described (Machado et al., 2003a; Oliveira et al., 2012). The best conditions for the reactions in laboratory scale were obtained at pH 6.80 using 33.00 mg L^{-1} of hydrogen peroxide for the degradation and mineralization of aqueous solutions containing 10 mg L^{-1} of paracetamol (França, 2011). Under these conditions, the substrate was completely oxidized after 40 minutes of reaction using TiO_2 P25, while 78% of mineralization with this same photocatalyst was reached after 120 minutes of reaction. Using the TiO_2/ZnPc composite, the substrate was completely oxidized after 60 minutes of reaction, and 63% was mineralized after 2 hours of reaction.

In the photocatalytic tests using a CPC reactor and solar radiation (**Fig. 8**), the experiments were done preferentially at pH 3.00 (França, 2011), using 50 L of an aqueous solution containing 10 mg L^{-1} of paracetamol and 100 mg L^{-1} of photocatalyst. Hydrogen peroxide, used as additional radical source (Machado et al., 2003a), was employed at the same concentration as defined in studies on laboratory scale.

Although the mineralization of paracetamol under the action of solar radiation has been equivalent in both cases (56%), after the accumulation of an UVA dose equal to 700 kJ m^{-2}, comparing the results obtained on laboratory scale and induced by solar radiation, it was observed that the increase in mineralization obtained with the use of the composite, 33%, was higher than that obtained using the commercial photocatalyst, equal to 25%, suggesting a better utilization of solar radiation by TiO_2/ZnPc composite.

In terms of degradation, monitored by high performance liquid chromatography (HPLC), the commercial photocatalyst required the accumulation of less UVA radiation (200 kJ m^{-2}) to oxidize 96% of paracetamol, whereas for the composite this level of degradation was achieved when the dose reached 350 kJ m^{-2} (**Fig. 9**).

The results obtained by Zhang et al (2010) indicated that TiO_2 photocatalytic degradation is an effective way to remove paracetamol from wastewater and drinking water without any generation of more toxic products. Although we have not analyzed the intermediates and products obtained, our results also point to the efficiency of heterogeneous photocatalysis in the treatment of acetaminophen, even if present in high concentrations in wastewater and drinking water.

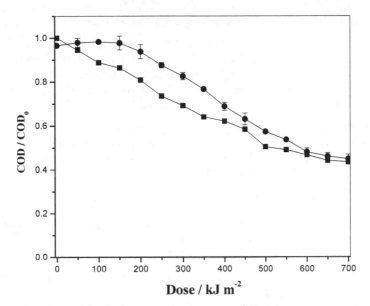

Fig. 8. Variation of dissolved organic carbon (DOC) as a function of cumulative dose of UVA, during the photocatalytic degradation of paracetamol mediated by solar radiation: $TiO_2/ZnPc$ 2.5% m/m (●);TiO_2 P25 (■).

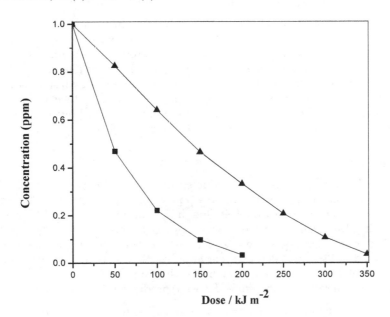

Fig. 9. Variation in the concentration of paracetamol measured by HPLC during photodegradation experiments mediated by TiO_2 P25 (■) and $TiO_2/ZnPc$ 2.5% m/m (▲) and induced by solar radiation.

3. Obtention of gaseous hydrogen for energy production

The International Energy Agency (IEA) estimates that world demand for energy should suffer an increase of 45% by 2030 (Birol, 2008). Based on the projections presented, one can expect a worsening of global warming, if no measures are taken that result in significant reduction of CO_2 emissions. In addition, we expect a worrying shortage of fossil fuels, if alternative sources of energy are not being widely used.

Among the alternative energy sources, H_2 is a very attractive option, as it concentrates high energy per unit mass – 1.0 kg of hydrogen contains approximately the same energy furnished by 2.7 kg of gasoline, which facilitates the portability of energy (Smith & Shantha, 2007). Besides, its combustion generates no contaminants.

Experts have pointed out three major obstacles to the expansion of consumption of hydrogen taking into consideration the technology available at the moment: clean production, low cost, storage and transportation. As a result, most efforts to expand the use of hydrogen as a source of cheap energy has been based on the development of new materials and processes of production.

Among the technologies for hydrogen production, biomass gasification (Albertazzi et al., 2005; Smith & Shantha, 2007), photocatalysis (Ni et al. 2007; Patsoura et al.,2007; Jing et al., 2010), and biological processes (Peixoto, 2008), have been focus of many studies for being routes clean and renewable. The heterogeneous photocatalysis and hydrogen generation by decomposition of water using concentrated solar radiation as primary source energy are between the most promising having gain attention due to their potential.

The great expectation of the global market for the use of hydrogen gas as an important input in the production of energy has been driven by the sectors of energy generation and distribution, which moves large numbers of capital around the world, and is in frank expansion, due to the enormous demand for energy by all sectors (Steinfeld, 2005; Preguer et al., 2009; Pagliaro et al., 2010). Most efforts to expand the use of hydrogen as a renewable energy source has been based on the development of fuel cell technology, both for expansion of its service life, by minimizing costs. Volumes of hydrogen gas have already been produced, both in EU-funded projects, such as the United States.

3.1 Hydrogen production using heterogeneous photocatalysis

In recent decades, research has been conducted on the possibility of using hydrogen as energy vector with low carbon emissions. The policy guidance for reducing the emission of greenhouse gases, and the prospect of decline in oil and other fossil fuels, has brought to light again the discussion about the use of hydrogen and technologies related to it. However, it is clear that large-scale use of hydrogen will only be possible if renewable sources are used in its production (Preguer et al., 2009). Currently, renewables contribute only about 5% of the commercial production of hydrogen, while the remaining 95% are derived from fossil fuels, given the still high cost of production from renewable sources.

The photocatalytic degradation of water to produce hydrogen, under the action of solar energy, offers a promising way to produce hydrogen cleanly, inexpensive and environmentally friendly. While great progress in photocatalysis using radiation in the ultraviolet region has occurred in recent decades, it has been extended with some difficulty, considering the use of visible radiation as a trigger for photocatalytic processes. Particularly, we have achieved some progress in this direction, involving the association between a photosensitizer dye and a semiconductor oxide.

The development of semiconductor oxides capable to be excited by radiation in the visible region became one of the most important topics in photocatalysis research, since the visible light represents a significant fraction of solar energy usable (Hwang et al., 2004). However, finding another photocatalyst than TiO_2, which has good chemical stability, corrosion resistance, be able to efficiently absorb radiation in the visible, and is environmentally friendly, has proved an arduous task. However, no semiconductor material capable of catalyzing the overall water splitting under action of visible radiation around 600 nm, with a quantum efficiency high enough to make possible the commercial application (Maeda & Domen, 2007; Jing et al., 2010). Besides, many of the photocatalysts capable to induce hydrogen production with commercially acceptable quantum efficiency, with excitation between 300 and 450 nm, are expensive and inadequate from the environmental point of view (Zeug et al., 1985; Maeda et al., 2006; Bao et al., 2008).

The low efficiency for the hydrogen production by semiconductor photocatalysis already with appropriate band gap should be due to the following reasons: 1) quick electron/hole recombination in the bulk or on the surface of semiconductor particles; 2) quick back reaction of oxygen and hydrogen to form water on the surface of catalyst; and 3) inability to promote efficient use of visible radiation. It is known that photogenerated electrons easily recombine with holes in the semiconductor (Hoffmann et al., 1995; Li et al., 2010; Kumar & Devi, 2011), compromising the quantum efficiency of the photocatalytic process (Kudo, 2006). Noble metal loading can suppress to some extent the charge recombination by forming a Schottky barrier (Chand & Bala, 2007; Fu et al., 2008). Often, sacrificial reagents has been added to the reaction media for the elimination of photo-generated holes, minimizing the electron/hole recombination, improving the quantum efficiency (Liu et al., 2006; Zaleska, 2008a; Jing et al., 2010). Methanol, ethanol and acetic acid have usually been employed as agents of sacrifice. Toxic organic substrates can also be a good option of sacrificial reagent (Jing et al., 2010).

Much progress has been made in photocatalytic water splitting since the Fujishima-Honda effect was reported (Fujishima & Honda, 1971, 1972). Thermodynamically, water splitting into H_2 and O_2 can be seen as an unfavorable reaction (ΔG = +238 kJ/mol) (Jing et al., 2010; Melo & Silva, 2011). However, the efficiency of water splitting is determined by the band gap, band structure of the semiconductor and the electron transfer process (Linsebigler et al., 1995; Hagfeldt & Grätzel, 1995; Melo & Silva, 2011).

Generally for efficient H_2 production using visible light-driven semiconductor the band gap should be less than 3.00 eV (ca. 420 nm) and higher than 1.23 eV (ca. 1000 nm), corresponding to the water splitting potential (Jing et al., 2010; Melo & Silva, 2011). Moreover CB and VB levels should satisfy the energy requirements set by the reduction and oxidation potentials for H_2O, respectively: the bottom of the conduction band must be located at a more negative potential than the H^+/H_2 reduction potential (Eo = 0 V $vs.$ NHE at pH 0), while the top of the valence band must be more positively positioned than the H_2O/O_2 oxidation potential (Eo = 1.23 V $vs.$ NHE) (Melo & Silva, 2011). Band engineering is thus necessary for the design of new semiconductors with the combined properties (Jing et al., 2010).

Oxides as $HPb_2Nb_3O_{10}$, $MgWO_x$ and $NiInTaO_4$ among others, active under the action of ultraviolet radiation, were also active in the visible region after doping using C, N and S (TiO_2N_x, TiO_2C_x, TaON and $Sm_2Ti_2O_5S$) (Hwang et al., 2004), as well as certain perovskite-type photocatalysts, with significant absorption in the visible. Zhang & Zhang (2009) reported the synthesis of a photocatalyst based on $BiVO_4$ which showed high photocatalytic activity in the visible region. However, most of these catalysts are not environmentally friendly as TiO_2.

Photocatalytic induced water-splitting technology involving nanosized TiO_2, despite the considerable variety of semiconductor photocatalysts capable to split water using solar energy and other photocatalytic processes has great potential to support an economy based on low-cost and environmentally friendly hydrogen production using solar radiation (Ashokkumar, 1998; Ni et al., 2007).

The photocatalytic hydrogen production using TiO_2 as photocatalyst can be schematized through **Figs. 1 and 10**.

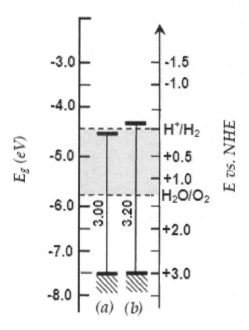

Fig. 10. Band gap of (a) Rutile sand (b) Anatase compared to the redox potential of water at pH 1.

For an efficient production of H_2, the energy level of the CB should be more negative than the energy level of the reduction of hydrogen, while the energy level of VB should be more positive than the energy level of the oxidation of water to formation of O_2 **(Fig.10)** (Ashokkumar, 1998; Ni et al., 2007), eqs 5 to 7. As outlined in **Fig. 1**, the vacancies photogenerated in the VB oxidize water into oxygen and hydrogen cations. These cations are reduced to molecular hydrogen in the conduction band. In other words, the separated electrons and holes act as reducer and oxidizer, respectively, in the water splitting reaction to produce hydrogen and oxygen. However, for this to happen effectively, it is necessary to ensure the fast transportation of the photogenerated carriers, avoiding bulk electron/hole recombination. Separation of hydrogen gas is also required as oxygen and hydrogen are produced simultaneously.

$$H_2O \xrightarrow{hv > E_g} H_2 + \frac{1}{2}O_2 \qquad (5)$$

$$2e^- + 2H^+ \xrightarrow{E_{BC} < E_{H^+/H_2O}} H_2 \qquad (6)$$

$$2h^+ + H_2O \xrightarrow{E_{BV} > E_{O_2/H_2O}} \frac{1}{2}O_2 + 2H^+ \qquad (7)$$

Having the adequate semiconductor, capable to induce water splitting when photoexcited by solar radiation, a key issue concerns the efficient utilization of the solar energy itself. Two major drawbacks of solar energy must be considered: (1) the intermittent and variable manner in which it arrives at the earth's surface (2) efficient collection of solar light on a useful scale. The first drawback can be solved by converting solar energy into storable hydrogen energy. For the second, the solution could be the use of solar concentrators (Jing et al., 2010).

For photocatalytic hydrogen production, it is imperative the use of visible radiation, especially if the goal is the storage of the energy supplied by the sun. Thus, photocatalysts able to mediate reactions through the use of visible radiation are more than desirable. Amplify the sensitivity of photocatalysts through the introduction of dopants, impurities and / or association between semiconductor and photosensitizers capable of shifting the absorption of the resulting composite to visible, are alternatives to a more efficient water photolysis (Hwang et al., 2004; Machado et al., 2008; Zaleska, 2008a, 2008b; Zhang & Zhang, 2009).

When a metal (eg platinum) is deposited on a semiconductor, the excited electrons migrate from the semiconductor to the metal until the Fermi levels of both species are aligned. The Schottky's barrier (Chand & Bala, 2007; Fu et al., 2008) formed at the metal/semiconductor interface can serve as a trap for electrons, efficient enough to minimize electron-hole recombination, increasing the efficiency of the photocatalytic process. At the same time, the metal is important for its own catalytic activity. Metals deposited on a semiconductor serve as active sites for the production of H_2, in which the trapped electrons are transferred to photogenerated protons to produce H_2 (**Fig. 11**) (Melo & Silva, 2011).

Research on photocatalytic hydrogen production in our laboratory is very recent. Our primary aim is the development of highly efficient, stable and low-cost visible-light-driven photocatalyst using different modification methods, such as doping, sensitization, supporting and coupling methods to extend the light response and performance of the photocatalyst aiming its application in environmental photocatalysis and photocatalytic hydrogen production. Despite a considerable variety of semiconductor photocatalysts capable to split water using solar energy and mediate other photocatalytic processes (Ashokkumar, 1998; Kim et al., 2010; Jing et al., 2010; Kumar & Devi, 2011), our studies have focused on improving the photocatalytic activity of TiO_2 through its synthesis by different procedures, their use and of hybrid variants, doped or not, and composites involving TiO_2 and photosensitizing dyes, especially considering issues related to the environment. In particular, we have studied photocatalytic reactions using solar radiation, with the photocatalyst in aqueous suspensions, with methodologies based on CPC reactor.

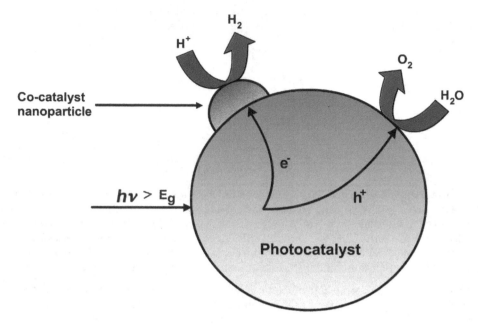

Fig. 11. Schematic representation of the photocatalytic water splitting on a platinized semiconductor powder particle.

We have developed small closed circulation reactor for bench-scale tests. These reactors ensure the evaluation of the developed photocatalyst from lab scale to out-door scale, in a batch mode.

The object of these studies is to improve hydrogen production and its storage under low pressure.

4. Solar cells

Photovoltaic cells are a good example of an alternative energy source, converting sunlight into electricity. Research in this field is quite intense given the importance of solar cells as sources of sustainable energy, as well as due to their reduced cost, low environmental impact, and fair efficiency for conversion of solar energy into electricity (O'Regan & Grätzel, 1991; Grätzel, 2003; Brennaman et al., 2011).

The efficiencies obtained for a silicon solar cell is about 24%, although at a very high manufacturing cost. Therefore, other materials have been studied in order to facilitate the conversion of solar energy into electrical energy (Zhao et al, 1998; Jayaweera et al., 2008; Cao et al., 2009; Patrocinio et al., 2010; Brennaman et al., 2011).

New developed devices such as dye solar cells, capable of converting solar energy into electrical (dye solar cells – DSCs), have been presented as alternatives for power generation (Hagfeldt & Gratzel, 1995; Gratzel & Hagfeld, 2000; Jayaweera et al., 2008). Despite its efficiency is still lower than that of silicon cells, the DSCs have been particularly interesting because they have a much lower production cost than devices based on silicon. In addition, the resulting material can be extremely thin as well as flexible and can be applied to almost any surface (Brabec et al., 2001; Kippelen & Bredas, 2009). Technically they are known as dye-sensitized solar cells, or DSSCs.

4.1 Dye-sensitized solar cells (DSSCs)

The use of solar cells based on a combination of dyes and photosensitizers oxide semiconductor (DSCs) have attracted great attention since the pioneering work of Grätzel and collaborators (O'Regan & Gratzel, 1991; Grätzel, 2005). The most efficient sensitizers for wide band gap semiconductors are the well-known metallo-organic ruthenium complexes (Grätzel & O'Regan, 1991). Certain Ru(II) complexes have shown to be excellent photosensitizers for TiO$_2$ in DSSCs, having gained the attention because of the high efficiencies achieved ($\eta \approx 11\%$) in converting sunlight into electricity (Nazeeruddin et al., 2005; Gao et al., 2008a, 2008b; Cao et al., 2009). In dye-sensitized solar cells, the conversion of visible light to electricity is achieved through the spectral sensitization of wide band gap semiconductors. Light is absorbed by the dye molecules, which are adsorbed on the surface of the semiconductor, thus inducing charge separation. Excitation of the dye molecules results in electron injection into the conduction band of the semiconductor. For electron injection to occur, the excited electrons must be at higher energy level than the semiconductor conduction band. An electrolyte of high ionic strength is also used in dye-sensitized solar cells to facilitate charge transfer across the device.

DSSCs have emerged as one of the most promising devices for sustainable photovoltaics due to their usually reduced cost, low environmental impact, and fair efficiency of conversion of solar energy into electricity (Grätzel, 2003; Polo et al., 2004; Brennaman et al., 2011).

Research in this field has been intense, given the growing worldwide demand for new energy sources (Kamat, 2007; Jacobson, 2009), either with focus on new materials and components or on cell assemblies for development of more efficient and environmentally friendly devices (Garcia et al., 2003; Prochazka et al., 2009; Zakeeruddin & Grätzel, 2009; Snaith, 2010). It is increasingly urgent the need to diversify energy matrices in order to rely on truly renewable energy sources, cleaner and environmentally friendly, if the goal is to build an ecologically sustainable society (Kamat, 2007; Herrero et al., 2011).

However, the high cost of dyes based on Ru (II), due to the low abundance of this metal and use restrictions from the environmental point of view are aspects that restricts its application on a large scale, which has stimulated efforts to use photosensitizing dyes with good features, safe and low cost (Hamann et al., 2008; Mishra et al 2009; Imahori et al., 2009).

Several simple organic dyes, and especially xanthene dyes (Eosin Y, Rose Bengal, etc.), yield efficiencies comparable to those achieved with ruthenium complexes, especially when used to sensitize ZnO films (Guillén et al., 2008; Plank et al., 2009; Pradhan et al., 2007). Organic dyes such as these are inexpensive (Kroon et al., 2007), can be easily recycled (Lee et al., 2006) and do not rely on the availability of precious metals such as ruthenium. They also have high extinction coefficients and their molecular structures contain adequate anchoring groups to be adsorbed onto the oxide surface. However, solar cells sensitized with such dyes tend to have low stability. The development and optimization of solar cells is of great interest, both commercially and scientifically. However, dye sensitized devices are still not commercially available in large volumes. Disadvantages such as the low efficiency and stability of these cells pose a hindrance to their commercialization.

A considerable increase in conversion efficiency of components of solar radiation into electrical energy by other photosensitizing dyes has been achieved in recent years. Macrocyclic systems such as porphyrins, phthalocyanines and derivatives have been shown to be capable of application in solar cells (Lu et al., 2009a).

Special attention has been given to the electron recombination processes that limit the DSC efficiency (Wang et al., 2006; Peter, 2007a; Zhao et al., 2008). Experimental and theoretical studies have been carried out in order to better understand and control these processes (Kruger et al., 2003; Cameron & Peter, 2005; Peter, 2007b; Xia et al., 2007a), typical interface phenomena. Strategies have been proposed to prepare efficient blocking layers in DSCs by using different techniques, such as spray pyrolysis, sputtering or by immersion in oxide precursor solutions (Xia et al., 2007a; Xia et al., 2007b; Wang et al., 2003; Handa et al., 2007). For example, the application of a compact layer onto the FTO glass before the mesoporous oxide film can prevent electron recombination at the FTO/TiO_2 interface. This blocking layer physically avoids the contact between the electrolyte and the FTO surface, decreasing the occurrence of triiodine reduction by photoinjected electrons (Patrocinio et al., 2010; Lei et al., 2010). Efficient layer-by-layer (LbL) TiO_2 compact films is considered one of the most effective blocking layers to avoid recombination processes at FTO surface in DSCs (Patrocinio et al.,. 2009). Although not previously reported as blocking layers, LbL metal oxide films have been applied in several devices (Krogman et al., 2008; Srivastava & Kotov, 2008; Jia et al., 2008; Lu et al., 2009b), including DSCs (Tsuge et al., 2006; Agrios et al., 2006). Iha and coworkers have shown that an LbL film based on TiO_2 nanoparticles and sodium sulphonated polystyrene, PSS, applied onto the FTO substrate before the mesoporous TiO_2 layer improved the overall conversion efficiency of DSCs by 28% (Patrocinio et al., 2009). Other complementary effects of the compact LbL TiO_2 layer in DSCs and the role of the polyelectrolyte itself were still under investigation.

LbL films using polyelectrolytes with good thermal stability at the electrode sintering temperature (450° C), such as sodium sulphonated polystyrene and sulphonated lignin, SL, maintain the compact morphology, and act as effective contact and blocking layers in DSCs. TiO_2 LbL films with poly(acrylic acid) as a polyanion presented similar morphology to that exhibited by TiO_2/PSS and TiO_2/SL films before sintering (Patrocinio et al., 2010). The best performance so far achieved is through the use of the TiO_2/PSS compact layer that increases the overall efficiency of DSCs to 30%, from 5.6 to 7.3%. The LbL TiO_2/PSS film imposes a longer time for a charge exchange at the electrode surface decreasing the electron recombination. The TiO_2/SL films (23% improvement) can be a cost effective option if a commercial application is considered.

5. Organic synthesis mediated by heterogeneous photocatalysis

Despite the widespread use of titanium dioxide, modified or not, or even other semiconductors with photocatalytic activity in photodegradation and mineralization of organic matter (Agostiano et al., 2003; Mrowetz et al.,2004; Machado et al.,2008; Hoffmann et al.,2010; Gupta et al.,2011), and its other capabilities (Mrowetz et. al.,2004; Zaleska, 2008a, 2008b), these semiconductors have been little explored in the synthesis of compounds of interest, although it is recognized that the photocatalytic synthesis should enable the efficient production of chemicals through combined fotoredox reactions with significant advantages compared to other methods (Swaminathan & Krishnakumar, 2011).

Synthetic methods based on photocatalytic processes have been reported for the preparation of different organic substrates (Amano et al., 2006; Palmisano et al., 2007b; Denmark & Venkatraman, 2006; Hakki et al.,2009; Swaminathan & Selvam, 2011; Swaminathan & Krishnakumar, 2011). Although the production of chemicals of industrial interest using heterogeneous photocatalysis has been shown to be a viable process, there is still little research on the use of photocatalysis for this purpose, and about the performance of these photocatalytic processes (Kanai et al., 2001; Murata et al., 2003; Amano et al., 2006; Denmark & Venkatraman, 2006; Hakki et al.,2009; Swaminathan & Selvam, 2011;Swaminathan & Krishnakumar, 2011). Apparently, the reason for this is that, in general, these methods are not yet fully satisfactory with regard to operational simplicity, cost of reagents and performance.

The stimulus for research in this field is necessary so that new and viable methodologies can be established.

6. Conclusion

In this chapter we combined a fast literature review about the different applications of heterogeneous photocatalysis, involving environmental photocatalysis, Hydrogen production for power generation, solar energy conversion into electricity using dye/semiconductor oxide cells and organic synthesis, with some experimental results obtained in our research group.

7. Acknowledgement

To Conselho Nacional de Desenvolvimento Científico e Tecnológico (CNPq), Coordenação de Aperfeiçoamento de Pessoal de Nível Superior (CAPES) and Fundação de Amparo à Pesquisa do Estado de Minas Gerais (FAPEMIG), Brazilian agencies for research funding and grants, and to Deutsches Zentrum für Luft- und Raumfahrt/Köln (DLR) and Bundesministerium für Bildung und Forschung (BMBF), Germany, by the support and funding.

8. References

Agostiano, A.; Cozzoli, P. D.; Comparelli, R.; Fanizza, E.; Curri, M. L. & Laub, D. (2004). Photocatalytic synthesis of silver nanoparticles stabilized by TiO_2 nanorods: A semiconductor/metal nanocomposite in homogeneous nonpolar solution. *Journal of the American Chemical Society*, Vol.126, No.12, (March 2004), pp. 3868-3879, ISSN 0002-7863

Agostiano, A.; Curri, M. L.; Comparelli, R.; Cozzoli, P. D. & Mascolo G. (2003). Colloidal oxide nanoparticles for the photocatalytic degradation of organic dye. *Materials Science & Engineering C-Biomimetic and Supramolecular Systems*, Vol.23, No.1-2, (January 2003), pp. 285-289, ISSN 0928-4931

Agrios, A. G.; Cesar, I.; Comte, P.; Nazeeruddin, M. K. & Grätzel, M. (2006). Nanostructured composite films for dye-sensitized solar cells by electrostatic layer-by-layer deposition. *Chemistry of Materials,*Vol.18, No.23, (November 2006), pp. 5395-5397, ISSN 0897-4756

Albertazzi, S.; Basile, E.; Brandin, J.; Einvall, J.; Hulteberg, C.; Fornasari, G.; Rosetti, V.; Sanati, M.; Trifiro, F. & Vaccari, A. (2005). The technical feasibility of biomass gasification for hydrogen production. *Catalysis Today*, Vol.106, No.1-4, (October 2005), pp. 297-300, ISSN 0920-5861

Amano, F.; Yamaguchi, T. & Tanaka, T. Photocatalytic Oxidation of Propylene with Molecular Oxygen over Highly Dispersed Titanium, Vanadium, and Chromium Oxides on Silica. *The Journal of Physical Chemistry B*, Vol.110, No.1, (January 2006), pp. 281-288, ISSN 1520-6106

Amat, A. M.; Bernabeu, A.; Vercher, R. F.; Santos-Juanes, L.; Simon, P. J.; Lardin, C.; Martinez, M. A.; Vicente, J. A.; Gonzalez, R.; Llosa, C. & Arques, A. (2011). Solar photocatalysis as a tertiary treatment to remove emerging pollutants from wastewater treatment plant effluents. *Catalysis Today*, Vol.161, No.1, (March 2011), pp. 235-240, ISSN 0920-5861

Andreozzi, R.; Caprio, V.; Insola, A. & Marotta, R. (1999). Advanced oxidation processes (AOP) for water purification and recovery. *Catalysis Today*, Vol.53, No.1, (October 1999), pp. 51-59, ISSN 0920-5861

Argazzi, R.; Bignozzi, C. A.; Hasselmann, G. M. & Meyer, G. J. (1998). Efficient Light-to-Electrical Energy Conversion with Dithiocarbamate-Ruthenium Polypyridyl Sensitizers. *Inorganic Chemistry*, Vol. 37, No.18, (September 1998), pp. 4533-4537, ISSN 0020-1669

Asbury,J. B.; Hao, E.; Wang, Y. Q.; Ghosh, H. N. & Lian, T. (2001). Ultrafast Electron Transfer Dynamics from Molecular Adsorbates to Semiconductor Nanocrystalline Thin Films. *Journal of Physical Chemistry B*, Vol. 105, No.20, (May 2001), pp. 4545-4557, ISSN 1089-5647

Ashokkumar, M. (1998). An overview on semiconductor particulate systems for production of hydrogen. *Internartional Journal of Hydrogen Energy*, Vol.23, No.6, (June 1998), pp. 427-438, ISSN 0360-3199

Augugliaro, V.; Palmisano, G.; Pagliaro, M. & Palmisano, L. (2007). Photocatalysis: a promising route for 21st century organic chemistry. *Chemical Communications*, Vol.33, No.33, (April 2007), pp. 3425-3437, ISSN 1359-7345

Bao, N. Z.; Shen, L. M.; Takata, T. & Domen, K. (2008). Self-templated synthesis of nanoporous CdS nanostructures for highly efficient photocatalytic hydrogen production under visible light. *Chemistry of Materials*, Vol.20, No.1, (January 2008), pp. 110-117, ISSN 0897-4756

Bartha, B.; Huber, C.; Harpaintner, R. & Schroder, P. (2010). Effects of acetaminophen in Brassica juncea L. Czern.: investigation of uptake, translocation, detoxification, and the induced defense pathways. *Environmental Science and Pollution Research*, Vol.17, No.9, (November 2010), pp. 1553-1562, ISSN 0944-1344

Batista, P. S. (2010). *Síntese e caracterização de novos fotocatalisadores de dióxido de titânio*. D.Sc. Thesis. Instituto de Química, Universidade Federal de Uberlândia, Uberlândia - MG, Brazil

Batista, P. S.; Santos, M. R. C.; de Souza, D. R.; Oliveira, D. F. M.; França, M. D.; Müller Jr, P. S. & Machado, A. E. H. (2011). Synthesis, characterization and photocatalytic activity of composites of TiO_2 and Zinc Phthalocyanine. X Brazilian Materials

Research Society Meeting. Symposium F: Nanostructured Functional Materials for Advanced Energy and Environmental Applications. Gramado, RS, Brazil. September, 2011

Benkö, G.; Kallioinen, J.; Tommola, J. E. I. K.; Yarstev, A. P. & Sundstrom, V. J. (2002). Photoinduced Ultrafast Dye-to-Semiconductor Electron Injection from Nonthermalized and Thermalized Donor States. *Journal of the American Chemical Society,*Vol.124, No.3, (January 2002), pp. 489-493, ISSN 1520-5126

Birol, F. (2008).*World Energy Outlook* London, UK: International Energy Agency (IEA)

Brabec, C. J.; Cravino, A.; Meissner, D.; Sariciftci, N. S.; Fromherz, T.; Rispens, M. T.; Sanchez, L. & Hummelen, J. C. (2001). Origin of the open circuit voltage of plastic solar cells. *Advanced Functional Materials*, Vol.11, No.5, (October 2001), pp. 374-380, ISSN 1616-301X

Bound, J. P. & Voulvoulis, N. Pharmaceuticals in the aquatic environment a comparison of risk assessment strategies. *Chemosphere*, Vol.56, No.11, (September 2004), pp. 1143–1155, ISSN 0045-6535

Brennaman, M. K.; Patrocinio, A. O. T.; Song, W.; Jurss, J. W.; Concepcion, J. J.; Hoertz, P. G.; Traub, M. C.; Murakami Iha, N. Y. & Meyer, T. J. (2011). Interfacial Electron Transfer Dynamics Following Laser Flash Photolysis of [Ru(bpy)$_2$((4,4'-PO$_3$H$_2$)$_2$bpy)]$^{2+}$ in TiO$_2$ Nanoparticle Films in Aqueous Environments, *ChemSusChem*, Vol.4, No.2, (February 2011), pp. 216 – 227, ISSN 1864-5631

Cameron, P. J. & Peter, L. M. (2005). How Does Back-Reaction at the Conducting Glass Substrate Influence the Dynamic Photovoltage Response of Nanocrystalline Dye-Sensitized Solar Cells? (2005), Vol.109, No.15, (March 2005), pp. 7392–7398, ISSN 1864-5631

Cao, Y. M.; Bai, Y.; Yu, Q. J.; Cheng, Y. M.; Liu, S.; Shi, D.; Gao, F.F. & Wang, P. (2009). Dye-Sensitized Solar Cells with a High Absorptivity Ruthenium Sensitizer Featuring a 2-(Hexylthio)thiophene Conjugated Bipyridine. *Journal of Physical Chemistry C*, Vol.113, No.15, (March 2009), pp. 6290-6297, ISSN 1932-7447

Carp, O.; Huisman, C. L. & Reller, A. (2004). Photoinduced reactivity of titanium dioxide. *Progress in Solid State Chemistry*, Vol.32, No.1-2, (November 2004), pp. 33–177, ISSN 0079-6786

Cavalheiro, A. A.; Bruno, J. C.; Saeki, M. J.; Valente, J. P. S. & Florentino, A. O. (2008). Effect of scandium on the structural and photocatalytic properties of titanium dioxide thin films. *Journal of Materials Science*, Vol.43, No.2, (January 2008), pp. 602-608, ISSN 0022-2461

Chand, S. & Bala, S. (2007). Simulation studies of current transport in metal-insulator-semiconductor Schottky barrier diodes. *Physica B-Condensed Matter*, Vol.390, No.1-2, (March 2007), pp. 179-184, ISSN 0921-4526

Chen, C.; Qi, X. & Zhou, B. (1997). Photosensitization of colloidal TiO$_2$ with a cyanine dye. *Journal of Photochemistry and Photobiology A: Chemistry*, Vol. 109, No.2, (September 1997), pp. 155-158, ISSN 1010-6030

Chen, Z.; Zhong, C.; Zhang, Z.; Li, Z.; Niu, L.; Bin, Y. & Zhang, F. (2008a). Photoresponsive J-Aggregation Behavior of a Novel Azobenzene–Phthalocyanine Dyad and Its Third-Order Optical Nonlinearity. *The Journal of Physical Chemistry B*, Vol. 112, No. 25, (June 2008), pp. 7387–7394, ISSN 1520-5207

Chen, C. Z.; Shi, J. Z.; Yu, H. J. & Zhang, S. J. (2008b). Application of magnetron sputtering for producing bioactive ceramic coatings on implant materials. *Bulletin of Materials Science*, Vol.31, No.6, (November 2008), pp. 877-884, ISSN 0250-4707

Corcoran, E.; Nellemann, C.; Baker, E.; Bos, R.; Osborn, D. & Savelli, H. (eds). 2010. *Sick Water? The central role of wastewater management in sustainable development.* A Rapid Response Assessment. United Nations Environment Programme,UN-HABITAT, GRID-Arendal. www.grida.no, ISBN: 978-82-7701-075-5, Norway

Daughton, C. G. & Ternes, T. A. (1999). Pharmaceuticals and personal care products in the environment: agents of subtle change. *Environmental Health Perspectives*, Vol.107, No.6 , (December, 1999), pp. 907–942, ISSN 0091-6765

Denmark, S. E. & Venkatraman, S. (2006). On the mechanism of the Skraup-Doebner-Von Miller quinoline synthesis. *Journal of Organic Chemistry*, Vol.71, No.4, (February 2006), pp. 1668-1676, ISSN 0022-3263

Diebold, U. (2003). The surface science of titanium dioxide surface. *Surface Science Reports*, Vol.48, No. 5-8, (January 2003), pp. 53 - 229, ISSN 0167-5729

Duarte, E. T. F. M., Xavier, T. P., De Souza, D. R., De Miranda, J. A., Machado, A. E. D., Jung, C., De Oliveira, L. & Sattler C. (2005). Construção e estudos de performance de um reator fotoquímico tipo CPC ("Compound Parabolic Concentrator). *Quimica Nova*, Vol.28, No.5,(Junho 2005), pp. 921-926, ISSN 0100-4042

Eguchi, K.; Fujii, H., Inata, K.; Ohtaki, M. & Arai, H. (2001). Synthesis of TiO_2/CdS nanocomposite via TiO_2 coating on CdS nanoparticles by compartmentalized hydrolysis of Ti alkoxide. *Journal of Materials Science*, Vol.36, No.2 (January 2001), pp.527-532, ISSN 0022-2461

Eisfeld, A. & Briggs J.S. (2006). The J- and H- bands of organic dye aggregates. *Chemical Physics*, Vol. 324, No. 2-3, (May 2006), pp. 376–384, ISSN 0301-0104

FDA, Compliance Program Guidance Manual, p. 10, http://www.fda.gov/downloads/Food/GuidanceComplianceRegulatoryInformation/ComplianceEnforcement/ucm073305.pdf

Fidder, H.; Knoester, J. & Wiersma, D. A. (1991). Optical properties of disordered molecular aggregates: A numerical study. *Journal of Chemical Physics*, Vol.95, No.11 (December 1991), pp. 7880-7891, ISSN 0021-9606

França, M. D. (2011). *Degradação de paracetamol empregando Tecnologia Oxidativa Avançada baseada em fotocatálise heterogênea, usando irradiação artificial e solar.* M.Sc. Dissertation. Universidade Federal de Uberlândia. Uberlândia, MG, Brazil

Fu, X.; Long, J.; Wang, X.; Leung, D. Y. C.; Ding, Z.; Wu, L.; Zhang, Z.; Li, Z. & Fu, X. (2008). Photocatalytic reforming of biomass: A systematic study of hydrogen evolution from glucose solution. *International Journal of Hydrogen Energy*, Vol. 33, No. 22, (November 2008), pp. 6484-6491, ISSN 0360-3199

Fujishima, A. & Honda K. (1971). Electrochemical evidence for mechanism of the primary stage of   photosynthesis. *Bulletin of the Chemical Society of Japan*, Vol.44, No. 4, (April 1971), pp. 1148–1150, ISSN 1348-0634

Fujishima, A. & Honda, K. (1972). Electrochemical Photolysis of Water at a Semiconductor Electrode. *Nature*, Vol.238, No.5358, (July 1972), pp. 37-38, ISSN 0028-836

Fujishima, A.; Zhang, X. & Tryk, D. A. (2007). Heterogeneous photocatalysis: From water photolysis to applications in environmental cleanup. *International Journal of Hydrogen Energy*, Vol.32, No.14, (September 2007), pp. 2664-2672, ISSN 0360-3199

Furube, A.; Asahi, T.; Masuhara, H.; Yamashita, H. & Anpo, M. (2001). Direct observation of a picosecond charge separation process in photoexcited platinum-loaded TiO$_2$ particles by femtosecond diffuse reflectance spectroscopy. *Chemical Physics Letters*, Vol.336, No.5-6, (September 2000), pp. 424-430, ISSN 0009-2614

Gao, F.; Wang, Y.; Shi, D.; Zhang, J.; Wang, M. K.; Jing, X. Y.; Humphry-Baker, R.; Wang, P., Zakeeruddin, S. M. & Grätzel, M. (2008a). Enhance the optical absorptivity of nanocrystalline TiO$_2$ film with high molar extinction coefficient ruthenium sensitizers for high performance dye-sensitized solar cells. *Journal of the American Chemical Society*, Vol.130, No.32, (July 2008), pp. 10720-10728, ISSN 0002-7863

Gao, F.; Wang, Y.; Zhang, J.; Shi, D.; Wang, M.; Humphry-Baker, R.; Wang, P.; Zakeeruddin, S. & Grätzel, M. (2008b). A new heteroleptic ruthenium sensitizer enhances the absorptivity of mesoporous titania film for a high efficiency dye-sensitized solar cell. *Chemical Communications*, Vol.23, No.23, (Juny 2008), pp. 2635–2637, ISSN 1359-7345

Garcia, C. G.; Polo, A. S. & Iha, N. Y. M. (2003). Fruit extracts and ruthenium polypyridinic dyes for sensitization of TiO$_2$ in photoelectrochemical solar cells. *Journal of Photochemistry and Photobiology a-Chemistry*, Vol.160, No.1-2, (August 2003), pp. 87–91, ISSN 1010-6030

Grätzel, M. & O'Regan, B. (1991). A Low-Cost, High-Efficiency Solar-Cell Based on Dye-Sensitized Colloidal TiO$_2$ Films. *Nature*, Vol.353, No.6346, (October 1991), pp. 737-740, ISSN 0028-0836

Grätzel, M. (2001). Photoelectrochemical cells. *Nature*, Vol.414, No.6861, pp. 338-344, ISSN 0028-0836

Grätzel, M. (2003). Dye-sensitized solar cells. *Journal of Photochemistry and Photobiology C-Photochemistry Reviews*, Vol.4, No.2, (October 2003), pp. 145–153, ISSN 1389-5567

Gratzel, M. (2005). Solar energy conversion by dye-sensitized photovoltaic cells. *Inorganic Chemistry*, Vol.44, No.20, (October 2005), pp. 6841-6851, ISSN 0020-1669

Guillén, E.; Casanueva, F.; Anta, J.; Vega-Poot, A.; Oskam, G.; Alcántara, R.; Fernández-Lorenzo, C. & Martín-Calleja, J. (2008). Photovoltaic performance of nanostructured zinc oxide sensitised with xanthenes dyes. *Journal of Photochemistry and Photobiology A: Chemistry*, Vol.200, No.2-3, (September 2008), pp. 364-370, ISSN 1010-6030

Gupta, V. K.; Jain, R.; Nayak, A.; Agarwal, S. & Shrivastava, M. (2011). Removal of the hazardous dye-Tartrazine by photodegradation on titanium dioxide surface. *Materials Science & Engineering C-Materials for Biological Applications*, Vol.31, No.5, (July 2011), pp. 1062-1067, ISSN 0928-4931

Hagfeldt, A. & Gratzel, M. (1995). Light-Induced redox reactions in nanocrystalline systems. *Chemical Reviews*, Vol.95, No.1, (January 1995), pp. 49-68, ISSN 0928-4931

Hagfeldt, A. & Gratzel, M. (2000). Molecular photovoltaics. *Accounts of Chemical Research*, Vol.33, No.5, (May 2000), pp. 269-277, ISSN 0001-4842

Hakki, A.; Dillert, R. & Bahnemann, D. (2009). Photocatalytic conversion of nitroaromatic compounds in the presence of TiO$_2$. *Catalysis Today*, Vol.144, No.1-2, (Juny 2009), pp. 154-159, ISSN 0920-5861

Hamann, T. W.; Jensen, R. A.; Martinson, A. B. F.; Ryswyk, H. V. & Hupp, J. T. (2008). Advancing beyond current generation dye-sensitized solar cells. *Energy & Environmental Science*, Vol.1, No.1, (Junu 2008), pp. 66-78, ISSN 1754-5692

Hanaor, D. A. H. & Sorrell C. C. (2011). Review of the anatase to rutile phase transformation. *Journal of Materials Science*, Vol.46, No.4, (February 2011), pp. 855-874, ISSN 0022-2461

Handa, S.; Haque, S. A. & Durrant, J. R. (2007). Saccharide Blocking Layers in Solid State Dye Sensitized Solar Cells.*Advanced Functional Materials*, Vol.17, No.15, (October 2007), pp. 2878-2883, ISSN 1616-3028

Hao, Y.; Yang, M.; Yu, C.; Cai, S.; Liu, M.; Fan, L. & Li, Y. (1998). Photoelectrochemical studies on acid-doped polyaniline as sensitizer for TiO_2 nanoporous film. *Solar Energy Materials & Solar Cells*, Vol.56, No.1, (January 1998), pp. 75-84, ISSN 0927-0248

Herrmann, J. M. & Guillard, C. (2002). New industrial titania photocatalysts for the solar detoxification of water containing various pollutants. *Applied Catalysis B: Environmental*, Vol.35, No.4, (January 2002), pp. 281-294, ISSN 0926-3373

Herrero, C.; Quaranta, A.; Leibl, W.; Rutherford, A. W. & Aukauloo, A. (2011). Artificial photosynthetic systems. Using light and water to provide electrons and protons for the synthesis of a fuel. *Energy & Environmental Science*, Vol.4, No.7, (February 2011), pp. 2353-2365, ISSN 1754-5692

Hoffmann, M. R.; Martin, T.; Choi W. & Bahnemann, D. W. (1995). Environmental applications of semiconductor photocatalysis. *Chemical Reviews*, Vol.95, No.1, (January 1995), pp. 69-96, ISSN 1520-6890

Hoffmann, M. (2000). Frenkel and charge-transfer excitons in quasi-one-dimensional molecular crystals with strong intermolecular overlap. in Doktor der Naturwissenchaften Dissertation; Technischen Universität Dresden, 2000

Hoffmann, M. R.; Choi J. & Park, H. (2010). Effects of Single Metal-Ion Doping on the Visible-Light Photoreactivity of TiO_2. *Journal of Physical Chemistry C*, Vol.114, No.2, (December 2010), pp. 783-792, ISSN 1932-7455

Huang, S.; Guo, X.; Huang, X.; Zhang, Q.; Sun, H.; Li, D.; Luo, Y. & Meng, Q. (2011). Highly efficient fibrous dye-sensitized solar cells based on TiO_2 nanotube arrays. *Nanotechnology*, Vol.22, No.31, (July 2011), pp. 315401- 315407, ISSN 1361-6528

Hwang, D. W.; Kim H. G.; Jang, J. S.; Bae, S. W.; Ji, S. M. & Lee, J. S. (2004). Photocatalytic decomposition of water-methanol solution over metal-doped layered perovskites under visible light irradiation. *Catalysis Today*, Vol.93-95, (September 2004), pp. 845-850, ISSN 0920-5861

Iliev, V.; Tomova, D.; Bilyarska, L.; Prahov, L. & Petrov, L. (2003). Phthalocyanine modified TiO_2 or WO_3-catalysts for photooxidation of sulfide and thiosulfate ions upon irradiation with visible light. *Journal of Photochemistry and Photobiology, A: Chemistry*, Vol.159, (July 2003), No. 3, pp. 281-287, ISSN 1010-6030

Imahori H.; Umeyama T. & Ito S. (2009). Large pi-Aromatic Molecules as Potential Sensitizers for Highly Efficient Dye-Sensitized Solar Cells. *Accounts of Chemical Research*, Vol.42, No.11, (November 2009), pp. 1809-1818, ISSN 1520-4928

Imhof, A. & Pine, D. J. (1997). Ordered macroporous materials by emulsion templating. *Nature*, Vol.389, No.6654, (October 1997), pp. 948-951, ISSN 0028-0836

Ino, D.; Watanabe, K.; Takagi, N. & Matsumoto, Y. (2005). Electron Transfer Dynamics from Organic Adsorbate to a Semiconductor Surface: Zinc Phthalocyanine on $TiO_2(110)$, *The Journal of Physical Chemisry B*,Vol.109, No.38, (June 2005), pp. 18018-18024, ISSN 1520-5207

Ismail, A. F.; Bolong, N.; Salim, M. R. & Matsuura, T. (2009). A review of the effects of emerging contaminants in wastewater and options for their removal. *Desalination*, Vol.239, No.1-3, (March 2009), pp. 229-246, ISSN 0011-9164

Iwaszuk, A. & Nolan, M. (2011). Charge compensation in trivalent cation doped bulk rutile TiO_2. *Journal of Physics: Condensed Matter*, Vol.23, No.334307,(August 2009), pp. 1-11, ISSN 0953-8984

Jacobson, M. Z. (2009). Review of solutions to global warming, air pollution, and energy security. *Energy & Environmental Science*, Vol.2, No.2, (December 2008), pp. 148–173, ISSN 1754-5692

Jayaweera, P. V. V.; Perera, A. G. U. & Tennakone, K. (2008). Why Gratzel's cell works so well. *Inorganica Chimica Acta*, Vol. 361, No.3, (February 2008), pp. 707-711, ISSN 0020-1693

Jia, Y.; Han, W.; Xiong, G. & Yang, W. (2008). Layer-by-layer assembly of TiO_2 colloids onto diatomite to build hierarchical porous materials, *Journal of Colloid and Interface Science*, Vol.323, No.2, (April 2008), pp. 326–331, ISSN 1095-7103

Jin, B.; Chong, M. N.; Chow, C. W. K. & Saint, C. (2010). Recent developments in photocatalytic water treatment technology: A review. *Water Research*, Vol. 44, No.10, (February 2010), pp. 2997-3027, ISSN 0043-1354

Jing, D., Guo, L., Zhao, L., Zhang, X., Liu, H., Li, M., Shen, S., Liu, G., Hu, X., Zhang, X., Zhang, K., Ma, L. & Guo, P. (2010). Efficient solar hydrogen production by photocatalytic water splitting: From fundamental study to pilot demonstration, *International Journal of Hydrogen Energy*, Vol.35, No.13, (July 2010), pp. 7087-7097, ISSN 0360-3199

Jirka, A. M. & Carter, M. J. (1975). Micro semi-automated analysis of surface and wastewaters for chemical oxygen demand. *Analytical Chemistry*, Vol.47, No.8, (July 1975), pp. 1397-1402, ISSN 0003-2700

Jones, O. A. H.; Voulvoulis, N. & Lester, J. N. Aquatic environmental assessment of the top 25 English prescription pharmaceuticals. *Water Research*, Vol.36, No.20, (December 2007), pp. 5013–5022, ISSN 0043-1354

Kamat, P.V. (2007). Meeting the Clean Energy Demand: Nanostructure Architectures for Solar Energy Conversion. *The Journal of Physical Chemistry C*, Vol. 111, (February 2007), No. 7, pp. 2834 - 2860, ISSN 1932-7455

Kanai, H.; Shono, M.; Hamada, K. & Imamura, S. (2001). Photooxidation of propylene with oxygen over TiO_2–SiO_2 composite oxides prepared by rapid hydrolysis. *Journal of Molecular Catalysis A:Chem*, Vol.172, No.1-2, (July 2001), pp. 25-31, ISSN 1381-1169

Khataee, A. R.; Zarei, M. & Ordikhani-Seyedlar, R. (2011). Heterogeneous photocatalysis of a dye solution using supported TiO_2 nanoparticles combined with homogeneous photoelectrochemical process: Molecular degradation products. *Journal of Molecular Catalysis a-Chemical*, Vol.338, No.1-2, (March 2011), pp. 84-91, ISSN 1381-1169

Kim, O.-K., Je, J., Jernigan, G., Buckley, L. & Whitten, D. (2006). Super-helix formation induced by cyanine J-aggregates onto random-coil carboxymethyl amylose as template. *Journal of the American Chemical Society*, Vol.128, No.2, (January 2006), pp. 510-516, ISSN 0002-7863

Kim, J. & Choi, W. (2010). Hydrogen producing water treatment through solar photocatalysis. *Energy & Environmental Science*, Vol.3, No.8, (May 2010), pp. 1042-1045, ISSN 1754-5692

Kippelen, B. & Bredas, J. L. (2009). Organic photovoltaics. *Energy & Environmental Science*, Vol.2, No.3, (December 2009), pp. 251-261, ISSN 1754-5692

Koca, A. & Sahin M. (2002). Photocatalytic hydrogen production by direct sun light from sulfide/sulfite solution. *International Journal of Hydrogen Energy*, Vol.27, No.4, (April 2002), pp. 363-367, ISSN 0360-3199

Köhler, J. & Schmid, D. (1996). Frenkel excitons in $NaNO_2$: excitation energy transfer and exciton coherence. *Journal of Physics: Condensed Matter*, Vol. 8, No. 2, (January 1996), pp. 115-141, ISSN 1361-648X

Krogman, K. C.; Zacharia, N. S.; Grillo, D. M. & Hammond, P. T. (2008). Photocatalytic layer-by-layer coatings for degradation of acutely toxic agents. *Chemistry of Materials*, Vol.20, No.5, (March 2008), pp. 1924–1930, ISSN 0897-4756

Kroon, J.; Bakker, N.; Smit, H.; Liska, P.; Thampi, K.; Wang, P.; Zakeerudin, S.; Grätzel, M.; Hinsch, A.; Hore, S.; Würfel, U.; Sastrawan, R.; Durrant, J.; Palomares, E.; Pettersson, H.; Gruszecki, T.; Walter, J.; Skupien, K. & Tulloch, G. (2007). Nanocrystalline dye-sensitized solar cells having maximum performance. *Progress in Photovoltaics*, Vol.15, No.1,(January 2007), pp. 1-18, ISSN 1062-7995

Krüger, J.; Bach, U. & Grätzel, M. (2001). High efficiency solid-state photovoltaic device due to inhibition of interface charge recombination. *Applied Physics Letters*, Vol.79, No.13, (September 2001), pp. 2085-2087, ISSN 0003-6951

Krüger, J.; Plass, R.; Grätzel, M.; Cameron, P. J. & Peter, L. M. (2003). Charge transport and back reaction in solid-state dye-sensitized solar cells: A study using intensity-modulated phorovoltage and photocurrent spectroscopy. *The Journal of Physical Chemistry B*, Vol.107, No.31, (August 2003), pp. 7536–7539, ISSN 1520-6106

Kudo, A. (2006). Development of photocatalyst materials for water splitting. *International Journal of Hydrogen Energy*, Vol.31, No.2, (February,2006), pp. 197–202, ISSN 0360-3199

Kumar, S. G. & Devi, L. G. (2011). A review on modified TiO_2 photocatalysis under VU/visible light: selected results and related mechanisms on interfacial charge carrier transfer dynamics. *The Journal of Physical Chemistry A*. DOI: 10.1021/jp204364a, ISSN 1089-5639

Labat, F.; Baranek, P. & Adamo, C (2008). Structural and eletronic properties of selected rutile and anatase TiO_2 surfaces: An ab initio investigation. *Journal of Chemical Theory and Computation*, Vol.4, No.2, (February 2008), pp. 341- 352, ISSN 1549-9618

Lee, W.; Okada, H.; Wakahara, A. & Yoshida, A. (2006). Structural and photoelectrochernical characteristics of nanocrystalline ZnO electrode with Eosin-Y. *Ceramics International*, Vol. 32, No. 5, (March 2006), pp. 495-498, ISSN 0272-8842

Lei, B. X.; Liao, J. Y.; Zhang, R.; Wang, J.; Su, C. Y. & Kuang, D. B. (2010). Ordered Crystalline TiO_2 Nanotube Arrays on Transparent FTO Glass for Efficient Dye-Sensitized Solar Cells. *The Journal of Physical Chemistry C*, Vol. 114, No.35, (December 2010), pp. 15228–15233, ISSN 1932-7455

Leznoff, C. C. & Lever, A. B. P. (1990). Phthalocyanines: Properties and Applications, VCH Publishers, ISBN 978-0-471-18720-2, New York, United States

Li, J.; Liu, Y.; Chen, X. & Burda, C. (2005). Photocatalytic degradation of azo dyes by nitrogen-doped TiO_2 nanocatalysts. *Chemosphere*, Vol.61, No.1, (March 2005), pp. 11-18, ISSN 0045-6535

Li, J.; Wang, C.; Yang, G. M.; Mele, G.; Slota, A R.; Broda, M. A.; Duan, M. Y.; Vasapollo, G.; Zhang, X. & Zhang, F. X. (2009). Novel meso-substituted porphyrins: Synthesis, characterization and photocatalytic activity of their TiO$_2$-based composites. *Dyes and Pigments*, Vol.80, No.3, (March 2009), pp. 321-328, ISSN 0143-7208

Linsebigler, A. L; Lu, G. Q. & Yates, J. T. (1995). Photocatalysis on TiO$_2$ surfaces: principles, mechanisms, and selected results. *Chemical Reviews*, Vol.95, No.3, (May 1995), pp. 735–758, ISSN 0009-2665

Liu, H.; Yuan, J. & Shangguan, W. F. (2006). Photochemical reduction and oxidation of water including sacrificial reagents and Pt/TiO$_2$ catalyst. *Energy & Fuels*, Vol.20, No.6, (November 2006), pp. 2289–2292. ISSN 1520-5029

Long, R.; Dai, Y,; Meng, G. & Huang, B. B. (2009). Energetic and electronic properties of X-(Si, Ge, Sn, Pb) doped TiO(2) from first-principle. *Physical Chemistry Chemical Physics*, Vol. 11, No. 37, (June 2009), pp. 8165-817, ISSN 1463-9076

Long, R. & English, N. J. (2011). Band gap engineering of double-cation-impurity-doped anatase-titania for visible-light photocatalysts: a hybrid density functional theory approach. *Physical Chemistry Chemical Physics*, Vol. 13, No. 30, (June 2011), pp. 13698-13703, ISSN 1463-9076

Lu, H. P.; Mai, C. L.; Tsia, C. Y.; Hsu, S. J.; Hsieh C. P.; Chiu C. L.; Yeh C. Y. & Diau, E. W. G. (2009a). Design and characterization of highly efficient porphyrin sensitizers for green see-through dye-sensitized solar cells. *Physical Chemistry Chemical Physics*, Vol.11, No.44, (September 2009), pp. 10270-10274, ISSN 1463-9076.

Lu, Y.; Lee, W. H.; Lee, H. S.; Jang, Y. & Cho, K. (2009b). Low-voltage organic transistors with titanium oxide/polystyrene bilayer dielectrics. *Applied Physics Letters*, Vol.94, No.11, (March 2009), pp. 1133031-1133033, ISSN 0003-6951

Luo, H. M.; Takata, T.; Lee, Y. G.; Zhao, J. F., Domen, K. & Yan, Y. S. Photocatalytic activity enhancing for titanium dioxide by co-doping with bromine and chlorine. *Chemistry of Materials*, Vol.16, No.5, (March 2004), pp. 846-849, ISSN 0897-4756

Machado, A. E. H.; de Miranda, J. A.; de Freitas, R. F.; Duarte, E. T. F. M.; Ferreira, L. F.; Albuquerque, Y. D. T.; Ruggiero, R.; Sattler, C. & de Oliveira, L. (2003a). Destruction of the organic matter present in effluent from a cellulose and paper industry using photocatalysis. *Journal of Photochemistry and Photobiology a-Chemistry*, Vol.155, No.1-3, (February 2003), pp. 231-241, ISSN 1010-6030.

Machado, A. E. H.; Miranda, J. A.; Sattler, C. & Oliveira, L. (2003b). Compósitos de ftalocianina de zinco e óxido de titânio, para emprego em processos fotocatalíticos e método para sua obtenção. Brazilian Patent, PI 03009203-3 A2, 2003.

Machado, A. E. H.; Xavier, T. P.; de Souza, D. R.; de Miranda, J. A.; Duarte, E. T. F. M.; Ruggiero, R.; de Oliveira, L. & Sattler, C. (2004). Solar photo-Fenton treatment of chip board production waste water. *Solar Energy*, Vol.77, No.5, (March 2004), pp. 583-589, ISSN 0038-092X.

Machado, A. E. H.; França, M. D.; Velani, V.; Magnino, G. A.; Velani, H. M. M.; Freitas, F. S.; Muller, P. S.; Sattler, C. & Schmucker, A. (2008). Characterization and evaluation of the efficiency of TiO$_2$/zinc phthalocyanine nanocomposites as photocatalysts for wastewater treatment using solar irradiation. *International Journal of Photoenergy*, Vol.2008, (March 2008), pp. 1-12, ISSN 1687-529X.

Machado, A. E. H., Menezes da Silva, V. H. & Ueno, L. T. (2011a). "Formation of Frenkel's J aggregates from zinc phthalocyanine: a m06 approach". XVI Simpósio Brasileiro de Química Teórica. (November 2011). Ouro Preto, MG, Brazil

Machado, A. E. H., França, M. D., Müller Jr, P. S., Borges, K.A. & dos Santos, L.M. (2011b). Synthesis of new photocatalysts based on TiO_2. Unpublished results.

Maeda, K.; Teramura, K.; Lu, D. L.; Takata, T.; Saito, N.; Inoue, Y. & Domen, K. (2006). Photocatalyst releasing hydrogen from water. *Nature*, Vol.440, No.7082, (March 2006), pp. 295-295, ISSN 0028-0836

Maeda, K. & Domen, K. (2007). New non-oxide photocatalysts designed for overall water splitting under visible light. *Journal of Physical Chemistry C*, Vol.111, No.22, (June 2007), pp. 7851-7861, ISSN 1932-7447

Malato, S.; Blanco, J.; Richter, C.; Curcó, D. & Giménez, J. (1997). Low-concentrating CPC collectors for photocatalytic water detoxification: comparison with a medium concentrating solar collector. *Water Science and Technology*, Vol.35, No.4, (January 1997), pp. 157–164, ISSN 0273-1223

Malato, S.; Blanco, J.; Vidal, A. & Richter, C. (2002). Photocatalysis with solar energy at a pilot-plant scale: an overview. *Applied Catalysis B*, Vol.37, No.1, (April 2002), pp. 1–15, ISSN 0926-3373

Martin, S. T.; Lee, A. T. & Hoffmann M. R. (1995). Chemical Mechanism of Inorganic Oxidants in the TiO2/Uv Process - Increased Rates of Degradation of Chlorinated Hydrocarbons. Environmental Science & Technology, Vol.29, No.10, (October 1995), pp. 2567-2573, ISSN 0013-936X

Meissner, D. & Rostalski, J. (2001). "Photovoltaics of interconnected networks," *Synthetic Metals*, Vol.121, No.1-3, (March 2001), pp. 1551–1552, ISSN 0379-6776

Melo, M. D. & Silva L. A. (2011). Photocatalytic Production of Hydrogen: an Innovative Use for Biomass Derivatives. *Journal of the Brazilian Chemical Society*, Vol.22, No.8, (August 2011), pp. 1399-1406, ISSN 0103-5053

Mi, J.; Guo, L.; Liu, Y.; Liu, W.; You, G. & Qian, S. (2003). Excited-state dynamics of magnesium phthalocyanine thin film. *Physics Letters A*, Vol.310, No.5-6, (April 2003), pp. 486-492, ISSN 0375-9601

Mills A.; Elliott N.; Parkin I. P.; O'Neill S. A. & Clark R. J. H. (2002). Novel TiO_2 CVD films for semiconductor photocatalysis. *Journal of Photochemistry and Photobiology a-Chemistry*, Vol.151, No.1-3, (August 2002), pp. 171-179, ISSN 1010-6030

Mills A. & Hunte S. L. (1997). An overview of semiconductor photocatalysis. *Journal of Photochemistry and Photobiology A:Chemistry*, Vol.108, No.1, (July 1997), pp. 1-35, ISSN 1010-6030

Miranda, J. A.; Machado, A. E. H. & Oliveira, C. A. (2002). Comparison of the photodynamic action of methylene blue and zinc phthalocyanine on TG-180 tumoral cells. *Journal of Porphirins and Phthalocianines*, Vol. 6, No. 1, (2002), pp. 43-49, ISSN 1088-4246

Mishra A.; Bauerle P. & Fischer M. K. R. (2009). Metal-Free Organic Dyes for Dye-Sensitized Solar Cells: From Structure: Property Relationships to Design Rules. *Angewandte Chemie-International Edition*, Vol.48, No.14, (March 2009), pp. 2474-2499, ISSN 1422-0067

Mrowetz, M.; Balcerski, W.; Colussi, A. J. & Hoffmann, M. R. (2004). Oxidative power of nitrogen-doped TiO_2 photocatalysts under visible illumination. *Journal of Physical Chemistry B*, Vol.108, No.45, (October 2004), pp. 17269-17273, ISSN 1520-6106

Murata, C.; Yoshida, H.; Kumagai, J. & Hattori, T. (2003). Active sites and active oxygen species for photocatalytic epoxidation of propene by molecular oxygen over TiO_2-SiO_2 binary oxides. *The Journal of Physical Chemistry B*, Vol.107, No.18, (May 2003), pp. 4364-4373, ISSN 1520-6106

Nazeeruddin, M. K.; Kay, A; Rodicio, I.; Humphry-Baker, R.; Müller, E.; Liska, P.; Vlachopoulos, N. & Grätzel, M. (1993). Conversion of light to electricity by cis-X2bis(2,2'-bipyridyl-4,4'-dicarboxylate)ruthenium(II) charge-transfer sensitizers (X = Cl-, Br-, I-, CN-, and SCN-) on nanocrystalline titanium dioxide electrodes. *Journal of the American Chemical Society*, Vol.115, No. 14, (July 1993), pp. 6382-6390, ISSN 0002-7863

Nazeeruddin, M. K.; Angelis, F. D.; Fantacci, S.; Selloni, A.; Viscardi, G.; Liska, P.; Ito, S.; Takeru, B. & Grätzel, M. (2005). Combined experimental and DFT-TDDFT computational study of photoelectrochemical cell ruthenium sensitizers. *Journal of the American Chemical Society*, Vol.127, No.48, (December 2005), pp. 16835-16847, ISSN 0002-7863

Ni, M.; Leung, M. K. H.; Leung, D. Y. C. & Sumathy, K. (2007). A review and recent developments in photocatalytic water-splitting using TiO_2 for hydrogen production. *Renewable & Sustainable Energy Reviews*, Vol.11, No.3, (April 2007), pp. 401-425, ISSN 1364-0321

Ni, J. R.; Xiong, L.; Sun, W. L.; Yang, Y. & Chen, C. (2011). Heterogeneous photocatalysis of methylene blue over titanate nanotubes: Effect of adsorption. *Journal of Colloid and Interface Science*, Vol.356, No.1, (April 2011), pp. 211-216, ISSN 0021-9797

Nikolaou, A.; Meric, S. & Fatta, D.(2007) Occurrence patterns of pharmaceuticals in water and wastewater environments. *Analytical Bioanalytical Chemistry*, Vol.387, No.4, (February 2007), pp. 1225-1234, ISSN 1618-2642

Nogueira R. F. P. & Jardim W. F. (1998). Fotocatálise heterogênea e sua aplicação ambiental. *Química Nova*, Vol.21, No.1, (July,1998), pp. 69-72, ISSN 0100-4042

Ohno, T., Mitsui, T. & Matsumura, M. (2003). Photocatalytic activity of S-doped TiO_2 photocatalyst under visible light. *Chemistry Letters*, Vol.32, No.4, (April 2003), pp. 364-365, ISSN 0366-7022

Oliveira, D. F. M.; Batista, P. S.; Muller Jr, P. S.; Velani, V.; França, M. D., de Souza, D. R. & Machado, A. E. H. (2012). Evaluating the effectiveness of photocatalysts based on titanium dioxide in the degradation of the dye Ponceau 4R. *Dyes and Pigments*, Vol. 92, No. 1, (January 2012), pp. 563-572, ISSN 0143-7208

O'regan, B. & Gratzel, M. (1991). A Low-Cost, High-Efficiency Solar-Cell Based on Dye-Sensitized Colloidal TiO_2 Films. *Nature*, Vol.353, No.6346, (October 1991), pp. 737-740, ISSN 0028-0836

Pagliaro, M.; Konstandopoulos, A. G.; Ciriminna, R. & Palmisano, G. (2010). Solar hydrogen: fuel of the near future. *Energy & Environmental Science*, Vol.3, No.3, (January 2010), pp. 279-287, ISSN 1754-5706

Palmisano, G.; Augugliaro, V.; Pagliaro, M. & Palmisano, L. (2007a). Photocatalysis: a promising route for 21st century organic chemistry. *Chemical Communications*, No.33, (April 2007), pp. 3425-3437, ISSN 1359-7345

Palmisano, L.; Palmisano, G.; Yurdakal, S.; Augugliaro, V. & Loddo, V. (2007b). Photocatalytic selective oxidation of 4-methoxybenzyl alcohol to aldehyde in aqueous suspension of home-prepared titanium dioxide catalyst. *Advanced Synthesis & Catalysis*, Vol.349, No.6, (April 2007), pp. 964-970, ISSN 1615-4150

Patrocinio, A. O. T.; Paterno, L. G. & Murakami Iha, N. Y. (2009). Layer-by-layer TiO(2) films as efficient blocking layers in dye-sensitized solar cells. *Journal of Photochemistry and Photobiology a-Chemistry*, Vol.205, No.1, (June 2009), pp. 23–27, ISSN 1010-6030

Patrocinio, A. O. T.; Paterno, L. G. & Murakami Iha, N. Y. (2010). Role of Polyelectrolyte for Layer-by-Layer Compact TiO2 Films in Efficiency Enhanced Dye-Sensitized Solar Cells. *The Journal of Physical Chemistry C*, Vol.114, No.41, (October 2010), pp. 17954–17959, ISSN 1932-7447

Patsoura, A.; Kondarides, D. I. & Verykios, X. E. (2007). Photocatalytic degradation of organic pollutants with simultaneous production of hydrogen. *Catalysis Today*, Vol.124, No.3-4, (June 2007), pp.94-102, ISSN 0920-5861

Peixoto G. (2008). *Produção de hidrogênio em reator anaeróbio de leito fixo ascendente a partir de água residuária de indústria de refrigerantes*. M.Sc. Dissertation. Universidade Federal de São Carlos. São Carlos , SP, Brazil

Peter, L. M. (2007a). Characterization and Modeling of Dye-Sensitized Solar Cells. *The Journal of Physical Chemistry C*, Vol. 111, No. 18, (May 2007), pp. 6601–6612, ISSN 1932-7447

Peter, L. (2007b). Transport, trapping and interfacial transfer of electrons in dye-sensitized nanocrystalline solar cells. *Journal of Electroanalytical Chemistry*, Vol.599, No.2, (January 2007), pp. 233-240, ISSN 0022-0728

Pichat, P.; Disdier, J.; Hoang-Van, C.; Mas, D.; Goutailler, G. & Gaysse, C. (2000). Purification/deodorization of indoor air and gaseous effluents by TiO_2 photocatalysis. *Catalysis Today*, Vol.63, No.2-4, (December 2000),pp. 363-369, ISSN 0920-5861

Plank, N.; Howard, I.; Rao, A.; Wilson, M.; Ducati, C.; Mane, R.; Bendall, J.; Louca, R.; Greenham, N.; Miura, H.; Friend, R.; Snaith, H. & Welland, M. (2009). Efficient ZnO Nanowire Solid-State Dye-Sensitized Solar Cells Using Organic Dyes and Core-shell Nanostructures. *The Journal of Physical Chemistry C.*, Vol.43, No.113, (October 2009), pp. 18515-18522, ISSN 1932-7447

Polo, A. S.; Itokazu, M. K. & Murakami Iha, N. Y. (2004). Metal complex sensitizers in dye-sensitized solar cells. *Coordination Chemistry Reviews*, Vol.248, No.13-14, (July 2004), pp. 1343–1361, ISSN 0010-8545

Pons, M. N.; Alinsafi, A.; Evenou, F.; Abdulkarim, E. M.; Zahraa, O.; Benhammou, A.; Yaacoubi, A. & Nejmeddine, A. (2007). Treatment of textile industry wastewater by supported photocatalysis. *Dyes and Pigments*, Vol.74, No.2, (April 2006), pp. 439-445, ISSN 0143-7208

Pradhan, B.; Batabyal, S. & Pal, A. (2007). Vertically aligned ZnO nanowire arrays in Rose Bengal-based dye-sensitized solar cells. *Solar Energy Materials & Solar Cells*, Vol.91, No.9, (May 2007), pp. 769-773, ISSN 0927-0248

Prashant V. K. (2007) Meeting the Clean Energy Demand: Nanostructure Architectures for Solar Energy Conversion. *J. Phys. Chem. C*, Vol.111, No.7, (February 2007), pp. 2834-2860, ISSN 1932-7447

Pregger, T.; Graf, D.; Krewitt, W.; Sattler, C.; Roeb, M. & Möller, S. (2009). Prospects of solar thermal hydrogen production processes. *International Journal of Hydrogen Energy*, Vol.34, No.10, (May 2009), pp.4256-4267, ISSN 0360-3199

Prochazka, J.; Kavan, L.; Zukalova, M.; Frank, O.; Kalbac, M.; Zukal, A.; Klementova, M.; Carbone, D. & Grätzel, M. (2009). Novel synthesis of the $TiO_2(B)$ multilayer templated films. *Journal of Materials Chemistry.*, Vol.21, No.8, (2009), pp. 1457–1464, ISSN 0897-4756

Qiu, X. H.; Nazin, G. V. & Ho, W. (2004). Mechanisms of reversible conformational transitions in a single molecule. *Physical Review Letters*, Vol.93, No.19, (November 2004), pp. 196806-1-196806-4, ISSN 1079-7114

Rehm, J. M.; McLendon, G. L.; Nagasawa, Y.; Yoshihara, K.; Moser, J. & Grätzel, M. (1996). Femtosecond Electron-Transfer Dynamics at a Sensitizing Dye-Semiconductor (TiO_2) Interface. *The Journal of Physical Chemistry*, Vol.100, No. 23 , (January 1996), pp. 9577-9578, ISSN 1932-7447

Ricciardi, G.; Rosa, A. & Baerends, E. J. (2001). Ground and Excited States of Zinc Phthalocyanine Studied by Density Functional Methods. *The Journal of Physical Chemistry A*, Vol. 105, No 21, (May 2001), pp. 5242-5254, ISBN 1520-5215

Sattler, C.; Oliveira, L.; Tzschirner, M. & Machado, A. E. H.(2004a). Solar photocatalytic water detoxification of paper mill effluents. *Energy*, V. 29, No. 5-6, (April-May 2004), pp. 835-843, ISSN 0360-5442

Sattler, C.; Funken, K. H.; Oliveira, L.; Tzschirner, M. & Machado, A. E. H. (2004b). Paper mill wastewater detoxification by solar photocatalysis. *Water Science and Technology*, V. 49, No. 4, pp. 189-193, ISSN 0273-1223

Senthilarasu, S.; Velumani, S.; Sathyamoorthy, R.; Subbarayan, A.; Ascencio, J.A.; Canizal, G.; Sebastian, P.J.; Chavez, J.A. & Perez, R. (2003). Characterization of zinc phthalocyanine (ZnPc) for photovoltaic applications. *Applied Physics A: Materials Science & Processing*, Vol. 77, No. 3-4, (August 2003), pp. 383–389, ISBN 1432-0630

Shang, J.; Zhao, F.; W., Zhu, T. & Li, J. (2011). Photocatalytic degradation of rhodamine B by dye-sensitized TiO_2 under visible-light irradiation. *Science China-Chemistry*, Vol.54, No.1, (January 2011), pp. 167-172, ISSN 1674-7291

Sharma, G. D.; Mathur, S. C. & Dube, D. C. (1991). Organic photovoltaic solar cells based on some pure and sensitized dyes. *Journal of Materials Science.*, Vol.26, No.24, (1991), pp. 6547-6552, ISSN 0022-2461

Sharma, G. D.; Kumar, R. & Roy, M. S. (2006). "Investigation of charge transport, photogenerated electron transfer and photovoltaic response of iron phthalocyanine (FePc):TiO_2 thin films," *Solar Energy Materials & Solar Cells*, Vol. 90, No. 1, (January 2006), pp. 32-45, 0927-0248

Shourong, Z.; Qingguo, H.; Jun, Z. & Bingkun, W. (1997). A study on dye photoremoval in TiO_2 suspension solution. *Journal of Photochemistry and Photobiology A: Chemistry*, Vol.108, No.2, (August 1997), pp. 235 – 238, ISSN 1010-6030

Smith, B. & Shanta, M. S. (2007). Membrane reactor based hydrogen separation from biomass gas – a review of technical advancements and prospects. *International Journal of Chemical Reactor Engineering*, Vol.5, (November 2007), pp. 1-10, ISSN 1542-6580

Snaith, H. J. (2010). Estimating the Maximum Attainable Efficiency in Dye-Sensitized Solar Cells. *Advenced Functional Materials*, Vol.20, No.1, (January 2010), pp. 13–19, ISSN 1616-3028

Spadavecchia, F.; Cappelletti, G.; Ardizzone, S.; Ceotto, M. & Falciola, L. (2011). Electronic Structure of Pure and N-Doped TiO_2 Nanocrystals by Electrochemical Experiments and First Principles Calculations. *The Journal of Physical Chemistry C*, 2011, Vol.115, No.14, (March 2011), pp. 6381-6391, ISSN 1932-7447

Srivastava, S. & Kotov, N. A. (2008). Composite Layer-by-Layer (LBL) Assembly with Inorganic Nanoparticles and Nanowires. *Accounts in Chemical Research*, Vol.41, No.12, (November 2008), pp. 1831–1841, ISSN 0001-4842

Steinfeld, A. (2005). Solar thermochemical production of hydrogen - a review. *Solar Energy*, Vol.78, No.5, (December 2005) pp. 603-615, ISSN 0038-092X

Stylidi, M.; Kondarides, D. I. & Verykios, X. E. (2004). Visible light-Induced Photocatalytic Degradation of Acid Orange 7 in Aqueous TiO_2 Suspensions. *Applied Catalysis B: Environmental*, Vol.47, No.3, (February 2004), pp. 189-201, ISSN 0926-3373

Swaminathan, M. & Krishnakumar, B. (2011). A convenient method for the N-formylation of amines at room temperature using TiO_2-P25 or sulfated titania. *Journal of Molecular Catalysis A-Chemical*, Vol.334, No.1-2, (January 2011), pp. 98-102, ISSN 1381-1169

Swaminathan, M. & Selvam, K. (2011). One-pot photocatalytic synthesis of quinaldines from nitroarenes with Au loaded TiO_2 nanoparticles. *Catalysis Communications*, Vol.12, No.6, (February 2011), pp. 389-393, ISSN 1566-7367.

Tennakone, K.; Kumarasinghe, A. P.; Kumara, G. R. R. A.; Wijayantha, K. G. U. & Sirimanne, P. M. (1997). Nanoporous TiO_2 photoanode sensitized with the flower pigment cyanidin. *Journal of Photochemistry and Photobiology A: Chemistry*, Vol.108, No.2-3, (August 1997), pp. 193-195, ISSN 1010-6030

Torres, T.; Claessens, C. G. & Hahn, U. (2008). Phthalocyanines: From outstanding electronic properties to emerging applications. *Chemical Record*, Vol.8, No.2, pp. 75-97, ISSN 1527-8999

Tsuge, Y.; Inokuchi, K.; Onozuka, K.; Shingo, O.; Sugi, S.; Yoshikawa, M. & Shiratori, S. (2006). Fabrication of porous TiO_2 films using a spongy replica prepared by layer-by-layerself-assembly method: Application to dye-sensitized solar cells. *Thin Solid Films*, Vol.499, No.1-2, (March 2006), pp. 396–401, ISSN 0040-6090

Wang, C. Y., Liu, C. Y., Wang, W. Q. & Shen, T. (1997). Photochemical events during the photosensitization of colloidal TiO_2 particles by a squaraine dye. *Journal of Photochemistry and Photobiology A: Chemistry*, Vol.109, No.2, (September 15 1997), pp. 159-164, ISSN 1010-6030

Wang, P., Zakeeruddin, S. M., Moser, J. E. & Grätzel, M. J. (2003). A new ionic liquid electrolyte enhances the conversion efficiency of dye-sensitized solar cells. *Journal of Physical Chemistry B*, Vol.107, No.48, (December 4 2003), pp. 13280–13285, ISSN 1520-6106

Wang, Z. S., Yanagida, M., Sayama, K. & Sugihara, H. (2006). Electronic-insulating coating of $CaCO3$ on TiO2 electrode in dye-sensitized solar cells: Improvement of electron lifetime and efficiency. *Chemistry of Materials*, Vol.18, No.12, (June 13 2006), pp. 2912–2916, ISSN 0897-4756

Wojnárovits, L., Palfi, T. & Takacs, E. (2007). Kinetics and mechanism of azo dye destruction in advanced oxidation processes. *Radiation Physics and Chemistry*, Vol.76, No.8-9, (August-September 2007), pp. 497-1501, ISSN 0969-806X

Wood, J. & Tauc, D. L. (1972). Weak Absorption Tails in Amorphous Semiconductors. *Physical Review B*, Vol.5, No.8, (April 1972); pp. 3144-3151, ISSN 0163-1829

Wu, T., Liu, G., Zhao, J., Hidaka, H. & Serpone, N. (1998). Photoassisted degradation of dye pollutants. V. Self-photosensitized oxidative transformation of Rhodamine B under visible light irradiation in aqueous TiO_2 dispersions. *Journal of Physical Chemistry B*, Vol.102, No.30, (July 23 1998), pp. 5845-5851, ISSN 1089-5647

Wu, T., Xu, S. J., Shen, J. Q., Chen, S., Zhang, M. H. & Shen, T. (2000). Photosensitization of TiO_2 colloid by hypocrellin B in ethanol. *Journal of Photochemistry and Photobiology a-Chemistry*, Vol.137, No.2-3, (December 2000), pp. 191-196, ISSN 1010-6030

Xagas, A. P., Bernard, M. C., Hugot-Le Goff, A., Spyrellis, N., Loizos, Z. & Falaras, P. (2000). Surface modification and photosensitisation of TiO_2 nanocrystalline films with ascorbic acid. *Journal of Photochemistry and Photobiology a-Chemistry*, Vol.132, No.1-2, (March 20 2000), pp. 115-120, ISSN 1010-6030

Xia, J. B., Masaki, N., Jiang, K. J. & Yanagida, S. (2007a). Sputtered Nb_2O_5 as an effective blocking layer at conducting glass and TiO_2 interfaces in ionic liquid-based dye-sensitized solar cells. *Chemical Communications*, No.2, pp. 138–140, ISSN 1359-7345

Xia, J. B., Masaki, N., Jiang, K. J. & Yanagida, S. (2007b). Sputtered Nb_2O_5 as a novel blocking layer at conducting Glass/TiO_2 interfaces in dye-sensitized ionic liquid solar cells. *Journal of Physical Chemistry C*, Vol.111, No.22, (June 7 2007), pp. 8092–8097, ISSN 1932-7447

Yamashita, H., Harada, M., Misaka, J., Takeuchi, M., Ichihashi, Y., Goto, F., Ishida, M., Sasaki, T. & Anpo, M. (2001). Application of ion beam techniques for preparation of metal ion-implanted TiO_2 thin film photocatalyst available under visible light irradiation: Metal ion-implantation and ionized cluster beam method. *Journal of Synchrotron Radiation*, Vol.8, (March 2001), pp. 569-571, ISSN 0909-0495

Yang, H. H., Guo, L. J., Yan, W. & Liu, H. T. (2006). A novel composite photocatalyst for water splitting hydrogen production. *Journal of Power Sources*, Vol.159, No.2, (September 22 2006), pp. 1305–1309, ISSN 0378-7753

Yang, K. S., Dai, Y., Huang, B. B. & Whangbo, M. H. (2008). Density Functional Characterization of the Band Edges, the Band Gap States, and the Preferred Doping Sites of Halogen-Doped TiO_2. *Chemistry of Materials*, Vol.20, No.20, (October 28 2008), pp. 6528-6534, ISSN 0897-4756

Zakeeruddin, S. M. & Grätzel, M. (2009). Solvent-Free Ionic Liquid Electrolytes for Mesoscopic Dye-Sensitized Solar Cells. *Advanced Functional Materials*, Vol.19, No.14, (July 24 2009), pp. 2187-2202, ISSN 1616-301X

Zaleska, A. (2008a). Characteristics of Doped-TiO_2 Photocatalysts. *Physicochemical Problems of Mineral Processing*, No.42, (July 2008), pp. 211-221, ISSN 1643-1049

Zaleska, A. (2008b). Doped-TiO_2: A Review. *Recent Patents on Engineering*, Vol.2, (July 2008), pp. 157-164, ISSN 1872-2121

Zaleska, A., Zielinska, A., Kowalska, E., Sobczak, J. W., Lacka, I., Gazda, M., Ohtani, B. & Hupka, J. (2010). Silver-doped TiO_2 prepared by microemulsion method: Surface properties, bio- and photoactivity. *Separation and Purification Technology*, Vol.72, No.3, (May 2010), pp. 309-318, ISSN 1383-5866

Zhang, X.; Wu, F.; Wu, X.; Chen, P. & Deng, N. (2008). Photodegradation of acetaminophen in TiO_2 suspended solution. *Journal of Hazardous Materials*, Vol.157, No.2-3, (September 2008), pp. 300 – 307. ISSN 0304-3894

Zeug, N., Bücheler, J. & Kisch, H. (1985). Catalytic formation of hydrogen and carbon-carbon bonds on illuminated zinc sulfide generated from zinc dithiolenes. *Journal of the American Chemical Society*, Vol.107, No.6, pp. 1459–1465, ISSN 0002-7863

Zhang, F., Zhao, J., Zang, L., Shen, T., Hidaka, H., Pelizzetti, E. & Serpone, N. (1998). TiO_2-assisted photodegradation of dye pollutants II. Adsorption and degradation kinetics of eosin in TiO_2 dispersions under visible light irradiation. *Applied Catalysis B-Environmental*, Vol.15, No.1-2, (January 15 1998), pp. 147-156, ISSN 0926-3373

Zhang, A. & Zhang, J. (2009). Characterization of visible-light-driven $BiVO_4$ photocatalysts synthesized via a surfactant-assisted hydrothermal method. *Spectrochimica Acta Part A*, Vol.73, (March 2009), pp. 336-341, ISSN 1386-1425

Zhang, H., Lv, X. J., Li, Y. M., Wang, Y. & Li, J. H. (2010). P25-Graphene Composite as a High Performance Photocatalyst. *ACS Nano*, Vol.4, No.1, (January 2010), pp. 380-386, ISSN 1936-0851

Zhao, J. H., Wang, A., Green, M. A. & Ferrazza, F. (1998). 19.8% efficient "honeycomb" textured multicrystalline and 24.4% monocrystalline silicon solar cells. *Applied Physics Letters*, Vol.73, No.14, (October 1998), pp. 1991-1993, ISSN 0003-6951

Zhao, Y., Zhai, J., He, J. L., Chen, X., Chen, L., Zhang, L. B., Tian, Y. X., Jiang, L. & Zhu, D. B. (2008). High-Performance All-Solid-State Dye-Sensitized Solar Cells Utilizing Imidazolium-Type Ionic Crystal as Charge Transfer Layer. *Chemistry of Materials*, Vol.20, No.19, (October 14 2008), pp. 6022-6028, ISSN 0897-4756

Zhou, Y., Dang, M., Li, H. & Lu, C. (2011). Preparation of high-aspect-ratio TiO_2 nanotube arrays and applications for dye-sensitized solar cells. *Optoelectronics and Advanced Materials-Rapid Communications*, Vol.5, No.5-6, (May 2011), pp. 523-526, ISSN 1842-6573

Ziolli, R. L. & Jardim, W. F. (1998). Mechanism reactions of photodegradation of organic compounds catalyzed by TiO_2. *Quimica Nova*, Vol.21, No.3, (May-June 1998), pp. 319-325, ISSN 0100-4042

5

Optimized Hybrid Modulation Algorithm to Control Large Unbalances in Voltage and Intensity in the NP Point of an NPC Converter

Manuel Gálvez, F. Javier Rodríguez and Emilio Bueno

Department of Electronics, Alcalá University

Spain

1. Introduction

The role of power-converters is growing rapidly in importance. In part, this is due to an increased use of renewable energy, which gets injected into the grid as electricity that must satisfy minimum quality regulations (Hammons 2011).

One of the features of using multilevel converters, as opposed to one level converters, is that they operate with different output voltage levels and as a result, a lower harmonic distortion coefficient is achieved. Moreover, these topologies allow working with higher voltages than the transistor break-down voltage, therefore allowing higher power ratings (Bueno 2005).

NPC multilevel converters can be connected to photovoltaic panels in such a way as to optimize the efficiency. This means that fewer panels need to connect directly to the grid, as opposed to two-level topologies, while peak line voltage does not exceed DC-bus voltage.

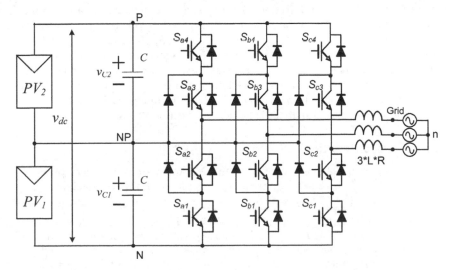

Fig. 1. Three-phase NPC converter, connected to a pair of independent photovoltaic-panel arrays.

There is however a drawback that comes up in the use of photovoltaic panels. Only low voltage can be supplied by these devices. This means that they must be connected in series, in order to achieve the desired voltage. Unfortunately, the lowest current supplying element sets the maximum current generated by the array.

Minimizing the number of panels connected in a series improves array efficiency once the converter is directly connected to a 400V grid and NP is connected between the two halves of the panel array. Furthermore, as these two arrays are independent, both can be set to inject maximum power at any moment.

Recently, there has been some work that has focused on transformerless photovoltaic inverters, and the influence of current leakage (Gonzalez et al., 2008), (Kerekes et al., 2009) (Kerekes et al., 2007). This is especially important for human saftey. The German standard, VDE0126-1-1, deals with grid-connected PV systems, and gives the requirements for limiting ground leakage and fault currents. These works coincide in that the NPC is an ideal topology for regulatory compliance, and the algorithm proposed minimizes the NP ripple voltage, one reason why the current leakage exists. The voltage level achieved by the capacitors is studied in several works, which focus on the NP point ripple (Bueno et al., 2006), (Celanovic & Boroyevich 2000), (Ogasawara & Akagi 1993) (Qiang et al., 2003).

(Pou et al., 2007) proposes to eliminate low frequency ripple at the NP using two modulations which contribute to an increase in switching frequency for transistors, and therefore power losses. Work by (Cobreces et al. 2006), based on a single-phase inverter, applies a changing state strategy in a specific duty cycle which also contributes to an increase in switching frequency.

Usually, there is no intention to tackle the asymmetric supply issue since it is very common to implement the same voltage for both capacitors thus avoiding an independent power supply implementation.

This document takes a look at this problem and tries to increase the efficiency of renewable energy generation. In Section 2, system characteristics are determined, and we introduce an improved proposal for (Pou et al., 2007) reducing NP ripple voltage in ideal conditions: plugging symmetric or minimal asymmetric power supply into the DC-bus. Section III explains the underlying principle of the proposed algorithm and implementation details. Section IV shows simulations made under "ideal" power supply conditions, and analyzes their limits, characteristics and advantages. In Section V, the power supply imbalance issue, due to the fact that power supplied by PV1 and PV2 are different, is discussed: the imbalance tolerance limit will be shown analytically. Finally, in Section VI, conclusions will be drawn.

2. System characteristics

Several different works have investigated the NP imbalance and have proposed several solutions. In (Pou et al., 2007), low frequency ripple is almost completely reduced at the expense of increasing switching frequency.

After having simulated the proposed grid-connected through an L filter with the DC-bus specifications listed in Table 1., Figures 2 and 3 show capacitor voltage evolution, NP ripple, inverter phase voltage and inverter phase-to-phase voltage. Figure 2 shows the results without the implementation of the improving algorithm, while Figure 3 shows the implementation of the algorithm proposed in (Pou et al., 2007).

Optimized Hybrid Modulation Algorithm to Control Large Unbalances in Voltage and Intensity in
the NP Point of an NPC Converter

101

I_{PV1}	I_{PV2}	U_{refVc1}	U_{refVc2}
100 A	100 A	375 V	375 V

Table 1. Simulation conditions.

Figure 2 demonstrates that NP ripple is considerable. This ripple is due to a low frequency component that matches with the third harmonic of the grid frequency, and a high frequency component that matches the switching frequency.

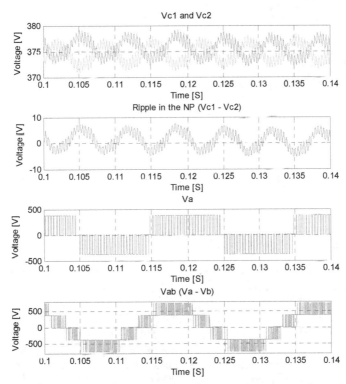

Fig. 2. Simulation results without the proposed algorithm. Capacitor voltage, NP ripple voltage, phase and line voltage.

Figure 3 show that the low frequency component is almost completely in the NP, but a high frequency component still exists.

Problems arise when the conditions in Table 1 change and voltage or current imbalances are introduced.

When this happens, and is due to small voltage imbalances, it is viable to introduce offsets and sort out the problem in both mentioned cases. An objection to this method is that modifying modulations may cause undesired over-modulation resulting in a discontinuous state. Moreover, falling into a discontinuous state becomes more likely when working close to the nominal power of the converter.

If this happens due to small current imbalances, it results in overcharging one capacitor, which brings an undesired voltage imbalance that can be solved with the same procedure, but with the same undesired consequences.

Therefore, connecting the NP to the middle point of a photovoltaic-panel array is an open field for investigation (Galvez et al., 2009) and (Busquets-Monge et al. 2008), and this technique is beginning to be applied to the connection of two wind turbine generators (Jayasinghe et al., 2010). However, currently, it is common to find arrays connected from UDC+ to UDC-. As shown in figure 4.

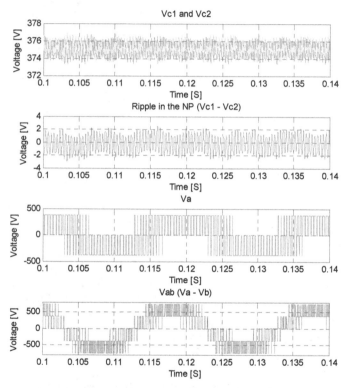

Fig. 3. Simulation results with algorithm proposed in (Pou et al., 2007). Capacitor voltage, NP ripple voltage, phase and line voltage.

The proposed algorithm solves this problem and improves performance even with bigger imbalances due to partial shading, a consequence of cloudy days and when some panels are dirty. Moreover the features of the panels, after manufacturing, are not identical for individual panels, and the photovoltaic panels performance changes over time are unequal.

What's more, there are some researchers who have proposed several topologies that work with more than one array of photovoltaic panels (Calais et al.,1998), and with special attention to the problem of current leakage (Gonzalez et al., 2008), (Kerekes et al., 2009) and (Kerekes et al., 2007).

Optimized Hybrid Modulation Algorithm to Control Large Unbalances in Voltage and Intensity in
the NP Point of an NPC Converter

103

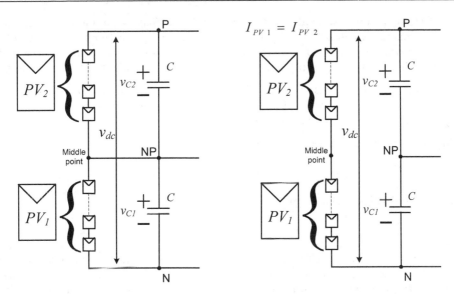

Fig. 4. Schematics of an array of photovoltaic panels. The left one with middle point
connected and the right one unconnected.

3. Underlying principle of the proposed algorithm

Regardless of the modulation technique applied, either SPWM or SVPWM (Bueno et al., 2002),
a set of states defines transistors in each phase. These states imply that each phase is connected
to 1, 0 or -1, where 1 is P (Udc+), 0 is NP (middle point) and -1 is N (Udc-). Because the
topology described in this document is triphasic, it is necessary to work with a three element
vector, where each value is the state of each phase, typical of a three-phase converter.

The aim of a converter is to fix a specific phase-to-phase voltage in order to force a current
flow through the grid filter. The system will work correctly as long as the appropriate
voltage is applied, whichever phase it was connected to 1, 0 or -1. Figure 5 shows the
schematic that works with two independent MPPTs, where the original vector is generated.
There has been a lot of work done on Maximum Power Points (MPPT) for photovoltaic
panels (Weidong et al., 2007), (Salas et al., 2006) and (Jain & Agarwal 2007), and in (Esram et
al., 2007) research focuses on the review of maximum power point tracking. In (Patel &
Agarwal 2008), the influence of shadows on MPPTs is studied.

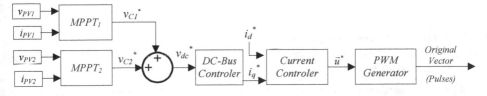

Fig. 5. Schematics of the control system.

The algorithm we propose explains how to calculate a fixed or dynamic voltage band around the reference voltage in NP, which can be Vdc/2 or something else. This voltage band, which will determine whether the measured NP voltage is above, below or within the range, will force the algorithm to increase, reduce or apply no offset to the NP voltage.

Once a reference shift is drawn, NP voltage must be changed, which means that one capacitor will charge as the other discharges with the same amount of energy. So, globally, energy and DC-bus voltage remain unchanged.

The instant power is distributed in the system in the following way:

$$P_G = P_{PV1} + P_{PV2} \tag{1}$$

where P_G is the total available or generated power,

$$p_{C1} + p_{C2} = P_{DC} \approx P_G - P_{Load} \tag{2}$$

and stored power in the capacitors is approximated, ignoring power losses in wires and the inverter. Nevertheless, this power can be considered more or less constant for transients, where $p_{C1} \neq p_{C2}$ is not only different, but not constant either; the variation being a consequence of the ripple in the NP. Hence, these powers can be separated into a continuous component and a variable component, as follows:

$$p_{C_1} = P_{C_1} + \frac{1}{2}C_1 \frac{du_{C_1}^{2}}{dt} \tag{3}$$

$$p_{C_2} = P_{C_2} + \frac{1}{2}C_2 \frac{du_{C_2}^{2}}{dt} \tag{4}$$

considering P_{DC} more or less constant, and existing for a short-period of time , we can say that:

$$\frac{1}{2}C_1 \frac{du_{C_1}^{2}}{dt} \approx -\frac{1}{2}C_2 \frac{du_{C_2}^{2}}{dt} \approx \Delta_p \tag{5}$$

and if (5) is substituted in (3) and (4), the equations become:

$$p_{C_1} \approx P_{C_1} + \Delta_p \tag{6}$$

$$p_{C_2} \approx P_{C_2} - \Delta_p \tag{7}$$

There are certain moments in which all phases are connected to the same capacitor, forcing that capacitor to get charged while the other one gets discharged. That is precisely the right time to change the normal behaviour.

The object is to modify the state vector in order to overrule the standard sequence and charge the appropriate capacitor.

Optimized Hybrid Modulation Algorithm to Control Large Unbalances in Voltage and Intensity in the NP Point of an NPC Converter

105

A list with all shifted vectors is shown in Table 2, but some conditions have to be met: the appropriate state combination and a true need of shift.

Upper shifts						Lower shifts					
Original Vector			Shifted Vector			Original Vector			Shifted Vector		
0	0	1	-1	-1	0	-1	-1	0	0	0	1
0	1	0	-1	0	-1	-1	0	-1	0	1	0
0	1	1	-1	0	0	-1	0	0	0	1	1
1	0	0	0	-1	-1	0	-1	-1	1	0	0
1	0	1	0	-1	0	0	-1	0	1	0	1
1	1	0	0	0	-1	0	0	-1	1	1	0

Table 2. Vector-Shift implementation.

The results found when a shifted vector is applied are that signs in (6) and (7) get changed. This forces a modified offset in the NP with twice the value and opposite sign. This is different from the original offset when shift is not done, giving:

$$p_{C_1}^* \approx p_{C_1} - 2\Delta_p \tag{8}$$

$$p_{C_2}^* \approx p_{C_2} + 2\Delta_p \tag{9}$$

and keeping the relationship in (2):

$$p_{C_1}^* + p_{C_2}^* = p_{C_1} + p_{C_2} = P_{DC} \tag{10}$$

Therefore, both aims of not modifying DC-bus voltage and quickly applying an appropriate change to the NP voltage are achieved.

Figure 6 shows where the change of the original vector for the shifted vector is put.

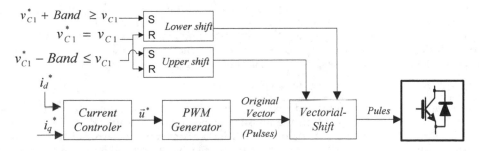

Fig. 6. Schematic with vector-shift implementation.

4. Characteristics of the proposed algorithm operation

The use of the proposed algorithm, not only achieves low frequency ripple reduction, but also cancels all of the components that make the NP voltage drop out of the limits of the desired band, making switching frequency dependent on the span of the ripple band.

First, we will start off with a system analysis performed under identical DC-bus conditions applied to Table 1.

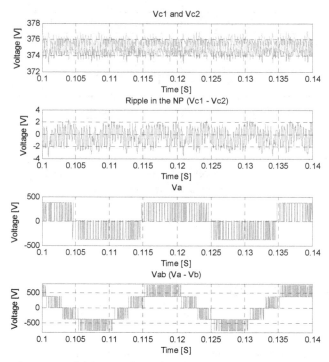

Fig. 7. Shows simulation results for a voltage band of ±1v.

Fig. 7 Simulation results with the proposed algorithm and ±1v band. C1 and C2 voltage ripple in the NP, phase voltage and line voltage.

Additionally, capacitor voltage could be checked to see if it is kept within the voltage range. It is possible to establish a narrow band in order to cancel even the high frequency components at the cost of increasing the switching frequency. So a decision must be made between ripple span or switching losses, considering the thermal limitation design to avoid any damage to the devices, and the transistors transient response.

Figure 8 shows simulation results for a voltage band of ±2v.

Moreover, it should be considered that DC-bus voltage is not truly constant, because the undesired ripple is dependent on the amount of power injected into the grid. Therefore, despite the fact that one capacitor benefits from a more limited narrow band, the other capacitor suffers from a wider band, the sum of that band and DC-bus ripple. That does useless to reduce ripple in the DC-bus.

Optimized Hybrid Modulation Algorithm to Control Large Unbalances in Voltage and Intensity in
the NP Point of an NPC Converter

107

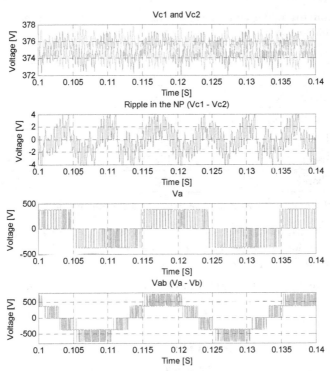

Fig. 8. Simulation results with the proposed algorithm and the ±2v band, C1 and C2 voltage ripple in the NP, phase voltage and line voltage.

This effect is shown in Figure 9. The parameters represented are C1 voltage; ripple voltage in the NP and DC-bus voltage. Also, the influence of a ±0.5V voltage band can be seen.

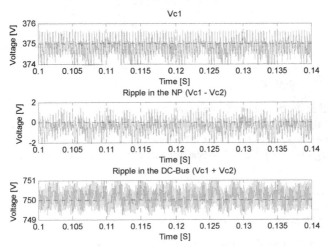

Fig. 9. Simulation results for C1 with ±0.5V ripple.

Table 3 shows both the amount and the rate of increase of switches for the applied algorithm with different voltage spans.

	Amount	Increase Rate %
Without Algorithm	606	0%
Proposed in (Pou et al., 2007)	845	39,44%
±2V band	761	25,58%
±1V band	1027	69,47%
±0,5V band	1749	188,61%

Table 3. Amount and rate of increase of switches in different simulations.

Therefore, it can be affirmed that reducing the ripple to NP ripple values causes an increase in switches and, therfore, losses, but does not reduce ripple in the DC-Bus. Because of this, it is not effective to use NP ripple values lower than DC-Bus ripple values.

So, when the power injected into grid is lower, the NP ripple is also lower and the number of times that the algorithm has to shift, is lower, making the number of swiches lower as well.

5. Characteristics of the proposed algorithm operation with high imbalances

The main feature of the algorithm is that it keeps the voltage in the NP controlled, even when there are big current imbalances during capacitor charge. This means that PV2 injects a different current than PV1.

Figure 10 shows the system simulation results with the proposed algorithm under the DC-bus specifications listed in Table 4, for a ±2V band.

I_{PV1}	I_{PV2}	U_{refVc1}	U_{refVc2}
125 A	75 A	375 V	375 V

Table 4. Simulations conditions.

It can be seen that both capacitor voltages remain within the fixed band, and that line voltage matches with the one simulated in the Figure 3. This, however, is not true for phase voltage since it can be clearly seen that it tends to be connected to +Udc for positive values, and to 0V for negative values. This does not happen when injected current into the bus is symmetric.

In this case the number of switches is 914, which means an increase rate of 50.82 % over the simulation without the algorithm. However, in this case there is no choice, as there is no way to keep bus balanced when such imbalances are applied.

As shown in Section II, the algorithm induces the vector shift at the moment that each phase is connected to the same capacitor, meaning that an adequate state combination happens. If

Optimized Hybrid Modulation Algorithm to Control Large Unbalances in Voltage and Intensity in
the NP Point of an NPC Converter

109

we consider that both capacitors have the same reference voltage, the maximum allowed imbalance can be calculated. If given:

$$D_V > 1 - m = 1 - \frac{U_{DC}}{\sqrt{2}V_L} \tag{11}$$

where D_V is the duty cycle when vector shift can be done, and m is the modulation index.

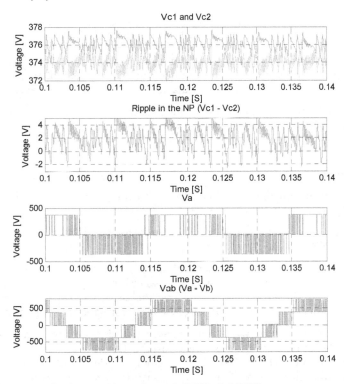

Fig. 10. Simulation results for a ±2V ripple with IPV2=0.6 ·IPV1.

Now, consider the following equations in terms of D_V:

$$P_{PV_{max}} = \frac{P_G}{2}(1 + D_V) \tag{12}$$

And, also:

$$P_{PV_{min}} = \frac{P_G}{2}(1 - D_V) \tag{13}$$

The Umb_{Max} is:

$$Umb_{Max} = \frac{P_{PV_{min}}}{P_{PV_{max}}} \tag{14}$$

From (12), (13) and (14):

$$\frac{P_{PV_{min}}}{P_{PV_{max}}} = \frac{(1-D_V)}{(1+D_V)} \tag{15}$$

Now, the maximum imbalance condition can be given, guaranteeing functionality of the algorithm. From (11), (12), (13), (14) and (15):

$$Umb_{Max} = \frac{(1-D_V)}{(1+D_V)} > \frac{U_{DC}}{\sqrt{8}V_L - U_{DC}} \tag{16}$$

Taking into account that V_L is considered constant, maximum asymmetry is fixed by DC-bus voltage, which is typically variable when coming from photovoltaic panels. Therefore, the system will work at its maximum performance as long as maximum asymmetry is met.

As a solution for higher asymmetries than the maximum asymmetry, the DC-bus voltage can be increased, although when this happens, it moves output away from the maximum power point of the panels.

6. Conclusion

A new hybrid modulation algorithm to control NP voltage has been described, even under high voltage asymmetries in the DC-bus.

It has been simulated both in symmetric and asymmetric power supplies, achieving voltage and ripple in the NP under control.

This algorithm is ideal for use in photovoltaic panel power source applications, as it tolerates high imbalances. If two independent panel arrays get connected, during cloudy days asymmetries will certainly occur. Nevertheless, each capacitor voltage can be controlled independently, greatly improving global performance of the panels.

Something else to keep in mind with photovoltaic panels is that fixing a capacitor voltage is a quick process, faster than the DC-bus time constant. This is one of the parameters that has the most influence on the changes in MPPT response under rapid perturbations from the environment. Furthermore, a reduction of one capacitor does not necessarily reduce DC-bus voltage; but reducing the degradation of the quality of injected energy in to grid by the fact that MPPT fixes the capacitor reference voltage. Moreover, this algorithm creates a new line of investigation for the design of faster MPPTs and faster responses for better THD.

The proposed algorithm has been validated with simulations shown in this document.

7. Acknowledgment

This work has been funded by the Spanish Ministry of Science and Education with reference number ENE2008-06588-C04-01.

8. References

Bueno, E. (2005). *Optimización del comportamiento de un convertidor de tres niveles NPC conectado a la red eléctrica*. Ph.D. thesis, Department of Electronics, University of Alcalá. http://www.depeca.uah.es/

Bueno, E.J., Cobreces, S., Rodriguez, F.J., Espinosa, F., Alonso, M. & Alcaraz, R. (2006).
    Calculation of the DC-bus Capacitors of the Back-to-back NPC Converters, Power
    Electronics and Motion Control Conference, 2006. EPE-PEMC 2006. 12th
    International, pp. 137-142, ISBN: 1-4244-0121-6

Bueno, E.J., Garcia, R., Marron, M., Urena, J. & Espinosa, F. (2002). *Modulation techniques
    comparison for three levels VSI converters*, Industrial Electronics Society, IEEE 2002
    28th Annual Conference 5-8 Nov. 2002, pp. 908 - 913 vol.2 ISBN: 0-7803-7474-6

Busquets-Monge, S., Rocabert, J., Rodriguez, P., Alepuz, S. & Bordonau, J.(2008). *Multilevel
    Diode-Clamped Converter for Photovoltaic Generators With Independent Voltage Control
    of Each Solar Array*, Industrial Electronics, IEEE Transactions on, Vol.55, pp. 2713-
    2723, ISSN: 0278-0046

Calais, M. & Agelidis, V.G. (1998). *Multilevel converters for single-phase grid connected
    photovoltaic systems-an overview*, Industrial Electronics, 1998. Proceedings. ISIE '98.
    IEEE International Symposium on, pp 224 - 229 vol.1 ISBN: 0-7803-4756-0

Celanovic, N. & Boroyevich, D. (2000). *A comprehensive study of neutral-point voltage balancing
    problem in three-level neutral-point-clamped voltage source PWM inverters*, Power
    Electronics, IEEE Transactions onVol.15, pp. 242-249, ISSN: 0885-8993

Cobreces, S., Bueno, E.J., Rodriguez, F.J., Salaet, J. & Bordonau, J. (2006). *A new neutral-point
    voltage control for single-phase tree-level NPC converters*, IEEE Power Electronics, pp.
    1-6, 18-22. ISSN : 0275-9306.

Esram, T. & Chapman, P.L. (2007). *Comparison of Photovoltaic Array Maximum Power Point
    Tracking Techniques Energy Conversion*, IEEE Transactions on, Vol. 22 pp. 439-449,
    ISSN: 0885-8969

Galvez, M., Bueno, E., Rodriguez, F.J., Meca, F.J. & Rodriguez, A. (2009). *New MPPT
    algorithm for photovoltaic systems connected to NPC converters and optimized for large
    variations of solar radiation*. Energy Conversion Congress and Exposition, 2009.
    ECCE 2009. IEEE pp. 48 - 53 ISBN: 978-1-4244-2893-9

Gonzalez, R., Gubia, E., Lopez, J. & Marroyo, L. (2008). *Transformerless Single-Phase
    Multilevel-Based Photovoltaic Inverter*, Industrial Electronics, IEEE Transactions on,
    Vol.55 pp. 2694-2702, ISSN: 0278-0046

Hammons, T. J. (2011). *Europe: Status of Integrating Renewable Electricity Production into the
    Grid, Electricity Infrastructures in the Global Marketplace*, ISBN: 978-953-307-155-8,
    InTech,   Available: http://www.intechopen.com/articles/show/title/europe-
    status-of-integrating-renewable-electricity-production-into-the-grid

Jain, S. & Agarwal, V. (2007). *Comparison of the performance of maximum power point tracking
    schemes applied to single-stage grid-connected photovoltaic systems*, Electric Power
    Applications, IET, Vol.1 pp. 753-762, ISSN: 1751-8660

Jayasinghe, S.D.G., Vilathgamuwa, D.M. & Madawala, U.K. (2010). *Connecting two wind
    turbine generators to the grid using only one three level NPC inverter* IECON 2010 - 36th
    Annual Conference on IEEE Industrial Electronics Society. pp 3263 - 3268 ISBN:
    978-1-4244-5225-5

Kerekes, T., Liserre, M., Teodorescu, R., Klumpner, C.& Sumner, M. (2009). *Evaluation of
    Three-Phase Transformerless Photovoltaic Inverter Topologies*, Power Electronics, IEEE
    Transactions on, Vol.24 pp. 2202-2211, ISSN: 0885-8993

Kerekes, T., Teodorescu, R. & Borup, U. (2007). *Transformerless Photovoltaic Inverters
    Connected to the Grid*, Applied Power Electronics Conference, APEC 2007 - Twenty
    Second Annual IEEE, pp. 1733 - 1737, ISBN: 1-4244-0713-3

Newton, C. & Sumner, M. (1997) *Neutral point control for multi-level inverters: theory, design and operational limitations.* Industry Applications Conference, 1997. Thirty-Second IAS Annual Meeting, IAS '97,5-9 Oct 1997, New Orleans, pp 1336 - 1343 vol.2 ISBN: 0-7803-4067-1

Ogasawara, S. & Akagi, H. (1993) *Analysis of variation of neutral point potential in neutral-point-clamped voltage source PWM inverters,* Industry Applications Society Annual Meeting, 1993.,pp. 965-970 vol.2, ISBN: 0-7803-1462-X

Patel, H. & Agarwal, V. (2008). Maximum Power Point Tracking Scheme for PV Systems Operating Under Partially Shaded Conditions, Industrial Electronics, IEEE Transactions on, vol.55 pp. 1689-1698, ISSN: 0278-0046

Pou, J., Zaragoza, J., Rodríguez, P., Ceballos, S., Sala, V.S., Burgos, R. P. & Boroyevich, D. (2007). *Fast-procesing modulation strategy for the neutral-point-clamped converter with total elimination of low-frequency voltage oscillations in de neutral point,* IEEE Trans. On Industrial Electronics, vol. 44, no. 4, pp. 2288-2294. ISBN: 0-7803-9252-3

Qiang, S., Wenhua, L., Qingguang, Y., Xiaorong, X. & Zhonghong, W. (2003) *A neutral-point potential balancing algorithm for three-level NPC inverters using analytically injected zero-sequence voltage,* Applied Power Electronics Conference and Exposition, 2003. APEC '03. pp. 228-233 vol.1, ISBN: 0-7803-7768-0

Salas, V., Olías, E., Barrado, A. & Lázaro, A. (2006). *Review of the maximum power point tracking algorithms for stand-alone photovoltaic systems,* Solar Energy Materials and Solar Cells, ELSEVIER, Vol.90, pp. 1555-1578

Weidong, X., Dunford, W.G., Palmer, P.R. & Capel, A. (2007). *Application of Centered Differentiation and Steepest Descent to Maximum Power Point Tracking,* Industrial Electronics, IEEE Transactions on, Vol.54 pp. 2539-2549, ISSN: 0278-0046

6

Utility Scale Solar Power with Minimal Energy Storage

Qi Luo and Kartik B. Ariyur

Purdue University
USA

1. Introduction

Smart grid functionality creating an internet of energy has been a topic of increasing interest. It is opening up several real time functions: pricing, network and consumption tracking, and integration of solar and wind power. The report from Department of Energy (DOE, 2008; 2009) supplies accessible details. At the present time, utilities run coal thermal power plants and nuclear plants as base load (Srivastava & Flueck, 2009) and use land based gas turbine plants to absorb unexpected demand surges (Nuqui, 2009). Solar energy, though envisioned as one of the panaceas to power from fossil fuels, suffers from two deficiencies: the density of energy available, and the unreliability of power production. Density limits mean solar power will never quite replace coal and nuclear plants for base load. However, the factor that limits penetration of solar power is the unreliability of supply that stems from uncertainty in incident solar radiation. A 100MW plant can produce much less power output in a matter of minutes if a cloud passes over it. It can also jump the other way. This can potentially result in large and undesirable transients being introduced into the grid. Large currents can damage grid equipment, such as power lines or transformers, in a very short period of time. This means that solar power needs backup power in the grid in the form of polluting coal or expensive gas. This is the main reason that grid operators and utilities are reluctant to integrate solar energy into their systems. This also means that solar or wind power at present may actually be contributing to greater use of fossil fuels in some regions. In this paper, we focus on solutions to three approaches to avoid the usage of grid storage: Distributing solar production to minimize its variance, correlation of solar power production to power consumption in air conditioning to determine the upper limits of solar penetration possible without storage, and grid failure probability with different levels of solar penetration.

The paper is organized as follows: Section 2 introduces the method of reducing supply uncertainty via geographic distribution of solar plants; Section 3 introduces the idea of matching solar output with air conditioning consumption and matching wind/solar power output with the consumption of electrical appliances; Section 4 demonstrates the probability of grid failure with different levels of solar penetration; Section 5 supplies concluding remarks.

2. Reducing supply uncertainty via geographic distribution

2.1 Solar power production

We consider a solar thermal system to produce electricity, in which the solar radiation is first absorbed by the receiver—a tube filled with working fluid (eg. molten salt, $150 - 350^{0}C$) and then the absorbed thermal energy is used as a heat source for a power generation system. Our analysis can be easily extended to PV systems – only the constants of proportionality will be different, yielding qualitatively similar results.

We assume flat plate solar collectors for analysis, which can be easily extended to cylindrical parabolic collectors (Singh & Shama, 2009).

$$P = A \cdot I \cdot \eta_1 \cdot \eta_2 \cdot r, \tag{1}$$

where A is the total area of collectors, I is the solar radiation intensity, and η_1 is energy transfer efficiency from solar radiation to thermal energy, η_2 is the Carnot Cycle energy transfer efficiency from thermal to mechanical energy, and r is the ratio of efficiency of real heat engine compared to the Carnot Cycle efficiency. We assume that conversion from mechanical to electrical energy is 100%.

The solar-thermal transfer efficiency η_1 can be calculated as:

$$\eta_1 = \tau\alpha - U_L \frac{T_H - T_a}{I}, \tag{2}$$

where τ is the transmissivity, α is the absorptivity listed in Tables 1 and 2. T_H is the average temperature of heat transfer fluid (usually melted salt or oil), and T_a is the ambient temperature.

Number of covers	$\tau\alpha$	$U_L (kW/m^2\ K)$
0	0.95	34
1	0.9	5.7
2	0.85	3.4

Table 1. Typical flat-plate solar collector (Black) properties

Number of covers	$\tau\alpha$	$U_L (kW/m^2\ K)$
0	0.90	28.5
1	0.85	2.8
2	0.80	1.7

Table 2. Typical flat-plate solar collector(Selective) properties

The Carnot efficiency η_2 in equation (1) is calculated :

$$\eta_2 = 1 - \frac{T_L}{T_H}, \tag{3}$$

where T_L is the lowest cycle temperature (which is slightly greater than ambient temperature T_a), T_H is the highest cycle temperature.

2.2 The idea of distributed solar power plant

Solar power plants can produce significant swings of power supply. A cloud passing over a 100MW plant can reduce its output to 20MW, and when it passes over, the output will again swing to 100MW. We develop here the idea of a distributed solar power plant which can ameliorate these swings. Similar work on distribution of wind plants has shown significant benefits (Archer & Jacobin, 2007). The difficulty here is that correlation of solar intensity in locations less than 100 miles from each other will make our quantitative results very different. In the analysis below we assume negligible correlation between solar intensity at multiple locations.

The construction of 10 plants of 10MW each will cause additional capital and maintenance costs. However these may be offset by the benefit of a steadier power supply and less damage to grid equipment. We show how this distribution may be systematically performed.

The power production of the distributed plant P_T is the sum of power produced in individual location $P_{i,}$:

$$P = \sum P_i, \tag{4}$$

The variance of power production in the distributed plant is

$$\sigma_p^2 = \sum_{i=1}^{n} f_i^2 \sigma_i^2, \tag{5}$$

where $f_i - \frac{P_i}{P_T}$, σ_i^2 is the variance of P_i, and

$$\sum_{i=1}^{n} f_i = 1, \tag{6}$$

σ_p^2 is minimized by the following solution:

$$f_i = \frac{1}{\sigma_i^2} \frac{1}{\sum_{i=1}^{n} \frac{1}{\sigma_i^2}}, \tag{7}$$

2.3 Hypothetical New York example

We use the historical data of solar intensity of twenty four candidate places within New York state from National Solar Radiation Data Base (NSRDB, 2005), and the corresponding temperature data from the United States Historical Climatology Network (USHCN, 2005). Then we choose four places of maximum annual solar intensity: Islip Long Island Macarthur Airport, John F Kennedy Intl Airport, New York Laguardia Airport and Republic Airport, and label them as area A, B, C and D. Fig. 1 shows the hourly average solar radiation of a typical day within each month. To construct synthetic time series of solar data, we proceeded as follows: Use random samples x_k from the data of solar intensity distribution between 2001-2005 every $T = 36$ seconds and use a low pass filter with a time constant $\tau = 180$ seconds. A valid question that may be asked here is – what is the benefit if solar intensity is strongly correlated between different locations? Our calculation of the covariance matrix of solar intensity for June, 2005 using hourly observations each day yields 30 difference covariance matrices for 30 days in June, 2005. The ratio of standard deviation to mean solar intensity of the corresponding eigenvalues of these 30 matrices range from 0.4681 to 0.8283 indicates a varying solar intensity distribution for different days. The ratio of the difference between eigenvalues and diagonal elements over diagonal elements range from -2.6496 to 0.9858 indicates a strong correlation of these four cites. However, in this discussion, we just neglect the correlation among these four cites. This opens up problems for future work which we discuss in the conclusions.

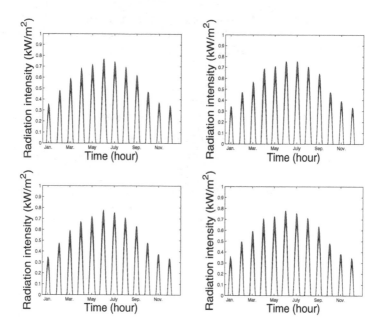

Fig. 1. Radiation intensity

The low pass filter smooths out jumps in intensity so they mimic what the motion of a cloud produces. The X axis in Fig. 1 is formed by one day from each month ($24hr \cdot 12month$).

The data of hourly average temperature of a typical day within each month is acquired from United States Historical Climatology Network (USHCN, 2005). Fig. 2, 3 and 4 show respectively the electric power outputs of the central plants in one location ($16000m^2 \times 1$), evenly in two locations ($8000m^2 \times 2$), and evenly in four locations ($4000m^2 \times 4$). Fig. 5 shows the power output of the optimally distributed plants. Fig. 6 gives the relationship between the coefficient of deviation and the installment cost. Y_1 axis represents the natural log of coefficient of deviation, Y_2 axis represents the natural log of setup cost. We can see from the figure that as the number of locations increases, the coefficient of deviation decreases, while the setup cost increases.

3. Supply-demand matching mechanisms

3.1 Matching solar production to air conditioner consumption

The electricity load in the hot season vs temperature for a large commercial facility in New York (Luo et al, 2009) in June, 2007 is shown in Fig. 7. The X axis is the temperature in ^0F, and Y axis is the electricity consumption in kWh. It is reasonable for the temperature and the load to have positive relation because in summer, the great portion of electricity consumption is due to air conditioning. Therefore the energy consumption of the building from the plot can be expressed as:

$$q_p = a_1 \times T + a_2 \tag{8}$$

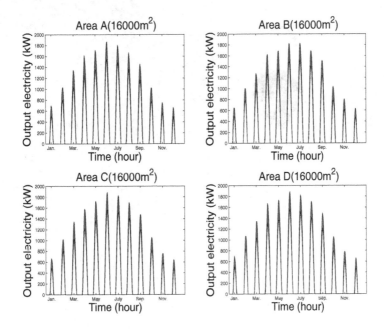

Fig. 2. Electricity output by CSP of area $16000m^2 \times 1$

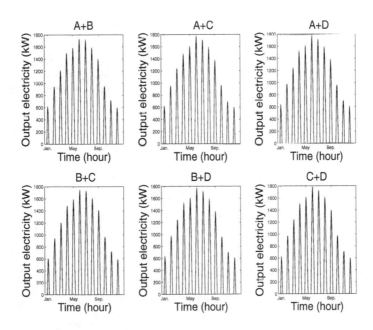

Fig. 3. Electricity output by CSP of area $8000m^2 \times 2$

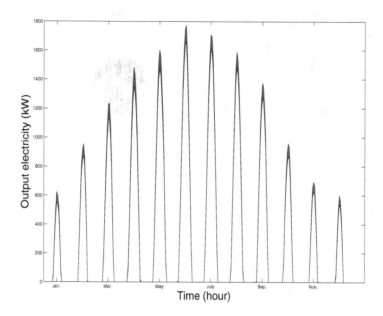

Fig. 4. Electricity output by CSP of area $4000m^2 \times 4$

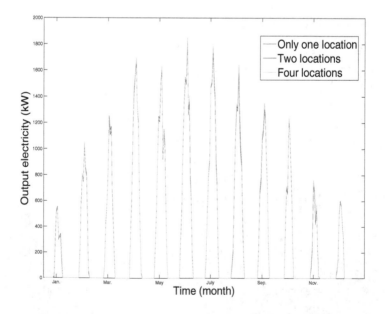

Fig. 5. Optimal electricity output by CSP of area located in four places

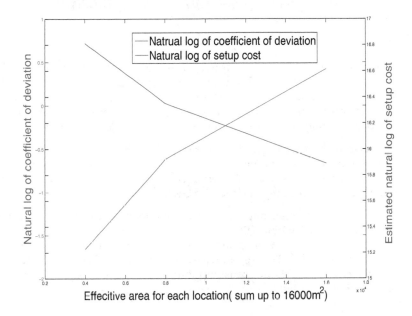

Fig. 6. Coefficient of deviation vs cost

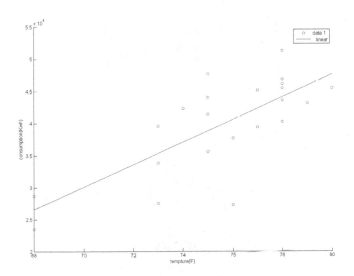

Fig. 7. Electricity consumption versus temperature in summer

$$C_p = \sum (a_1 \times T + a_2 - D_s)P_e + CSP_f + \sum D_s CSP_v, \qquad (9)$$

$$a_1 = 1756.9, a_2 = -92880 \qquad (10)$$

Where q_p is the predicted energy consumption, C_p is the predicted bill, D_s is the solar energy output, P_e is the electricity price, CSP_f is the fixed CSP maintenance cost, and CSP_v is the CSP cost that may vary according to the solar energy output. The reason the correlation of power consumption to temperature is not very strong in Fig. 7 is that, for the commercial facility , air conditioning consumption is a large but not the dominant part of consumption.

This idea comes from the simple fact that as the solar intensity increases, both the CSP output and the air conditioner consumption increase, so we can match them to achieve an energy balance. The advantage of this matching may include: reduce the need for base load plants, and integrating solar power stably into the grid base.

Fig. 8 gives the electricity consumption of air conditioner of 10000 families (with room area uniformly distributed within $80m^2 - 160m^2$ and power of air conditioner uniformly distributed within $0.8kW - 1.6kW$). It is calculated using active energy management described in previous work (Luo et al, 2009). We can see by comparing Fig. 5 and Fig. 8 that in summer, they have similar envelopes. The AC consumption flattens out because of the on-off nature of the control through thermostats.

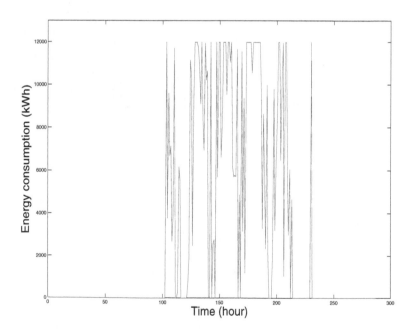

Fig. 8. AC consumption of 10000 families

Appliance	Power(kW)	Hours	Energy(kJ)
Water heater (40 gallon)	5	7	35
Clothes Dryer	4.5	2	9
Dishwasher	0.2	3.5	7
Hair dryer	1.6	5	8
Clothes iron	1.6	2	3.2
Vacuum cleaner	1.3	2	2.6
Toaster	1.3	3	3.9
Coffee maker	1.1	2	2.2
Refrigerator	0.7	50	35
Personal computer	0.27	32	8.6
Televisions	0.15	20	3

Table 3. Weekly energy consumption of home appliances

3.2 Household appliance consumption periodicity

Similar ideas can work for matching electronics appliances consumption and wind or solar energy output. Energy consumption of most appliances like dish washer/dryer have relatively constant frequencies and phases—although these may vary from family to family—since most people have the habit of washing their dishes and clothes at regular time of every week. This gives us the idea to describe the energy consumption of different electronics appliances with pulses of different frequencies and phases uniformly distributed over their respective ranges.

For a typical American family of four people, we have the room properties as in Table 3. In Fig. 9 we have shown the energy consumption distribution. We represent the energy consumption of the electronics applicants of each family with rectangle pulse train, and Fig. 10 gives the combination results of energy consumption of 100, 000 families.

So long as the total production from wind or solar as the renewable source equals consumption from the appliances within the period of consideration—such as a few hours or a day— it is theoretically possible, via pricing, to match supply and demand.

4. Grid dynamics and control with solar power penetration

4.1 Grid dynamics and control model

Traditional grid dynamics models have been discussed in various books (Murty, 2008), (Machowski et al, 2008). The flow chart 11 below gives the Gauss-Seidel iterative method for load flow solutions for a n bus system with 1 slack bus. In this flow chart, P is the real power in kW while Q is reactive power in kVar. In real system, in order to secure the grid system, we need constrains in this dynamics model such as:

- Current in any of n buses must not exceeds the limit I_{limit} in order to prevent blackout in the system.

- Calculated reactive power must not exceed the limits that local reactive power station can provide. If so, reactive power is fixed at the limit that is violated and it is no longer possible to hold desired bus voltage.

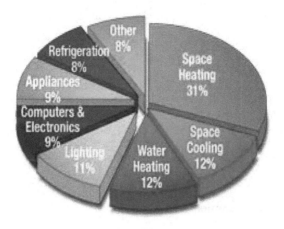

Fig. 9. Residential electricity consumption

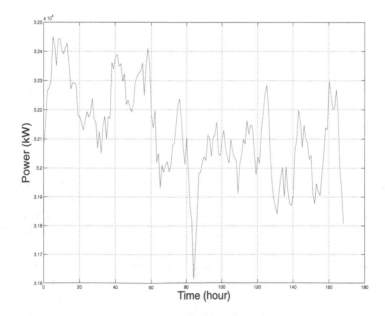

Fig. 10. Residential power consumption of 10000 families

Action/Operation	Time Frame
Wave effects(fast dynamics, lighting-caused overvoltages)	Microseconds to milliseconds
Switching overvoltages	Milliseconds
Fault protection	100 ms
Electromagnetic effects in machine windings	Milliseconds to seconds
Stability	1 seconds
Stability augmentation	Seconds
Electromechanical effects of oscillations in motors & generators	Milliseconds to minutes
Tie line load frequency control	1-10 seconds, ongoing
Economic load dispatch	10 seconds-1 hour, ongoing
Thermodynamic changes from boiler control action	Seconds to hours
System structure monitoring	1 hr- 1 day
System state estimation	1-10 seconds
Security monitoring	1 minute to 1 hour
Load management, forecasting	1 hour to 1 day, ongoing
Maintenance scheduling	Months to 1 year, ongoing
Expansion planning	Years, ongoing
Power plant building	2-10 years, ongoing

Table 4. Control Time Scales (Abdallah, 2009)

- The state of change of the voltage must not exceed the slow state of the generator+tie line voltage control. Otherwise we have control saturation and the generator can no longer track the changes.

Table 4 gives the time scale of different disturbances and control signals in the grid. Once the disturbance introduced to the grid system exceeds the control limit, the cascading failure will happen (Ding et al, 2011). The test case (Dusko et al, 2006) in Fig. 12 and Fig. 13 represent a large European power system and has 1000 buses, 1800 transmission lines and transformers, and 150 generating units. The base case load has an active power demand of 33 GW and a reactive power demand of 2.5 GVar. It is seen from 12 that there is a sharp increase in blackout size at the critical loading of 1.94 times the base case loading, and (Dusko et al, 2006) also discussed that the expected energy not served (EENS) share similar distribution for critical loading and under critical loading cases, and for over critical loading cases, we have different patterns, with the exponent of the power law distributions ranging from -1.2 to -1.5 as shown in Table. 5.

4.2 Grid fluctuation introduced by different levels of solar penetration

A report published in 2009 by the North American Electric Reliability Corporation showed that the output power of a large PV systems, with ratings in the order of tens of megawatts, can change by ±70% in a five- to ten-min time frame (NAERC, 2009). And it should also be mentioned that if a number of small systems that are distributed over a large land area, the resulting combined fluctuations are much less due to the smoothing effect according to our previous analysis.

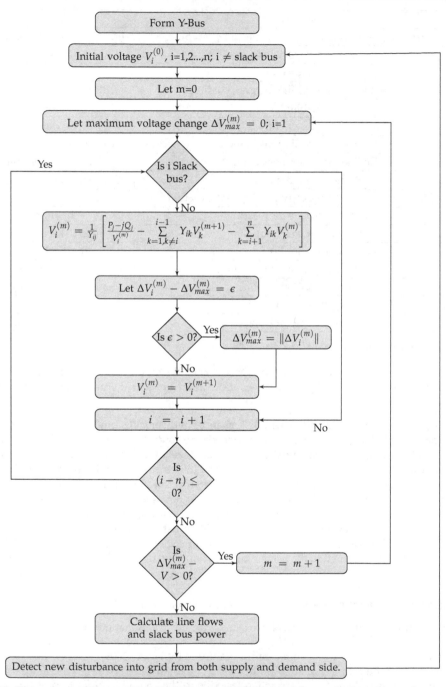

Fig. 11. Flow chart of Gauss-Seidel iterative method for load flow solutions for a n bus system with 1 slack bus

Model	Exponent	Test Case	Reference
OPA	-1.2, -1.6	-	(Carreras et al, 2004)
Branching	-1.5	-	(Dobson et al, 2004)
CASCADE	-1.4	1000 buses	(Dobson et al, 2005)
Hidden Failure	-1.6	-	(Chen et al, 2005)
Manchester	-1.2, -1.5	-	(Dusko et al, 2006)

Table 5. Approximate power law exponents at criticality for several cascading failure models

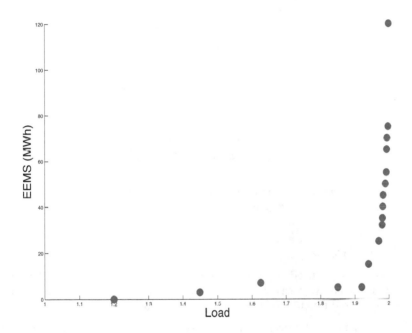

Fig. 12. Expected energy not served (EENS) as a function of the loading factor with respect to the base case. (Dusko et al, 2006)

Fig. 14 shows an example of the PV output fluctuations in New York area, it is a normal distribution with a mean plot and a confidence interval of 68%($\pm 1\sigma$). Here we assume the PV output fluctuation have similar distribution with the solar radiation, which is reasonable according to the solar power output model (Dusabe et al, 2009). The X axis of the figure is time in hours and Y axis is the system output in MWh. We can conclude from this figure that the most severe fluctuation occurs around noon. In general, the change of solar power output is usually due to:

• Time of the day.

• Time of the year.

• PV system locations.

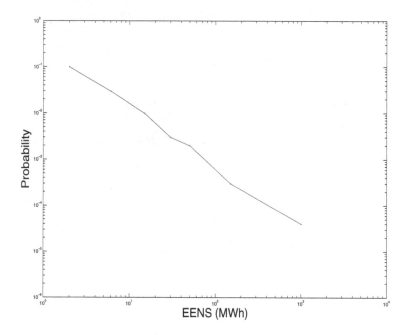

Fig. 13. Probability distribution of expected energy not served (EENS) at the critical loading of 1.94 times the base case loading (Dusko et al, 2006)

- Types of Clouds.
- PV system topology.

The negative effects introduced, especially to the stability of grid system as solar penetration level increases, is a major concern for the future grid. We can calculate the blackout probability of power system with different level of solar penetration as following:

For a grid system with $a\%$ of power from solar system, which follows a normal distribution:

$$W_{solar} \sim N(a\%, \sigma^2_{(a,i)}), i = 0, 1...23 \qquad (11)$$

where $a\%$ is the normalized expected power output from PV system at i^{th} hour of the day with $a\%$ of penetration for the overall grid system, and $\sigma_{(a,i)}$ is the standard deviation of power output at the same time and same penetration level.

We can equate the solar system fluctuation to the inverse change of loading, for instance, a decrease of $1MWh$ of solar production is equivalent to an increase of the power load at the same time frame, therefore the equivalent load should also follow normal distribution.

$$L \sim N(l, \sigma^2_{(a,i)}) \qquad (12)$$

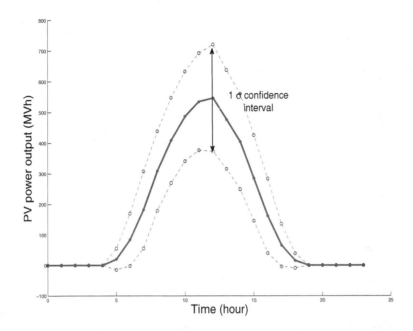

Fig. 14. Fluctuations in the output power of a large PV system (1σ confidence interval) (NSRDB, 2005)

then following the same procedure as in Fig. 12 and Fig. 13, probability of failure due to different level of solar penetration is

$$P_{(f,a)} = \int_{-\infty}^{\infty} p_{(L,a)} F_{(EENS>0|(L,a))} dL \tag{13}$$

where the probability $F_{(EENS>0|(L,a)}$ is the cumulative probability distribution of system failure

$$F_{(EENS>0|(L,a))} = \int_{0}^{\infty} (p_{(EENS,L,a)}) dEENS \tag{14}$$

and $p_{(EENS,L,a)}$ is the probability density function in Fig. 13, in this example, $L = 1.94$ and $a = 0$.

For instance, if we choose $i = 12$, when solar radiation follows N(547,174), and assume solar system output follows the same distribution. The normalized solar penetration of $a = 1, 10, 50, 100$, and corresponding standard deviation of solar system output $\sigma_{(a,i)} = 0.0027$, 0.0269, 0.1344, 0.2687. And the grid failure model with these levels of solar penetration is shown in Fig. 15, in which we show that as the level of solar penetration increases, the probability of system failure increases. This analysis does not take into account, additional solar backup. Thus as the proportion of solar power increases, the proportion of a controllable base load power source to meet demand fluctuations reduces, and hence, the system becomes more prone to failure. The availability of storage can ameliorate the problem. Fig. . 15 does not take into account real time matching of AC demand with solar supply, which and reduce failure probability.

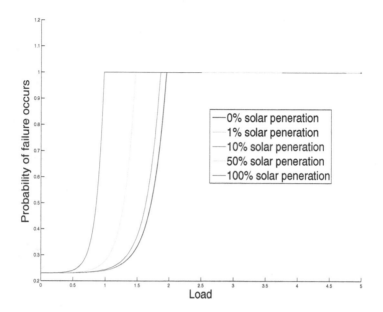

Fig. 15. Probability of system failure with different levels of solar penetration

5. Concluding remarks

We have supplied a partial proof of concept for two methods of reducing supply-demand imbalance. First, we have shown that the effect on power production of distributing solar plants - the coefficient of deviation is reduced by 50% in our example. If we were to consider covariance of energy production on different locations, our optimization problem for distribution could be extremely complicated, given that the covariance varies from day to day and from season to season. We will address this in future work, as the issue will obscure presentation of the basic idea here. Our calculations show that solar production and AC power consumption are strongly correlated. Hence, solar production could penetrate the grid to the extent of replacing the peaker gas turbine plants that the utilities use for peak usage in summer. Finally, we have shown by use modeling of aggregate demand of household appliances that it is possible to use solar or wind power as it gets produced. Whenever there is availability of solar power on the grid, smart appliances can switch on and use it.

Our ultimate objective is to reduce the unpredictability of supply-demand in the grid as solar power penetration increases, with minimal use of expensive, limited life grid storage. While the solution approaches we have proposed show how this can be done in theory, they ignore the transients that depend upon the speed of sensing supply and matching it with demand. The main requirement for stability in the grid is the matching of phase from various sources over a reactive time scale and load scheduling over the tactical time scale (minutes to hours). High speed measurements are available for both grid voltage and current. The movement of clouds is also reasonably predictable in short interval of less than hours. This will ensure that a utility can fire up a base load coal or steam turbine 3 hours before it is necessary or a gas turbine 15 minutes before it is necessary.

The utility industry is extremely conservative and will not make changes that can destabilize the grid. Even in Germany, solar penetration has not exceeded 2% inspect of significant taxpayer subsidies. What we have shown is that specific guarantees of safety can be constructed for various levels of solar penetration–whether distributed or centralized. Once we construct these guarantees, grid penetration of solar power could perhaps reach 10% even without advances in battery or thermal storage technology.

6. References

Abdallah, C.T.(2009). *Electric Grid Control: Algorithms & Open Problems*, available at `http://ElectricGridControl:Algorithms&OpenProblems`.

Archer, C. and Jacobin, M. Z.(2007). Supplying Baseload Power and Reducing Transmission Requirement by Interconnecting Wind Farms, *Journal of Applied Meteorology and Climatology* Volume 46.

Carreras BA, Lynch VE, Dobson I, Newman DE(2004). Complex dynamics of blackouts in power transmission systems. *Chaos;* 14(3), 43-52.

Chen J, Thorp JS, Dobson I. Cascading dynamics and mitigation assessment in power system disturbances via a hidden failure model. *International Journal of Electrical Power & Energy Systems*27(4):318-26.

Ding L., Cao Y., Wang W., Liu M.(2011), Dynamical model and analysis of cascading failures on the complex power grids. *Kybernetes*, Vol. 40 Issue 5, 814-823.

Dobson I, Carreras BA, Newman DE (2004). A branching process approximation to cascading load-dependent system failure. *Hawaii International Conference on System Sciences*, Hawaii, USA.

Dobson I, Carreras BA, Newman DE(2005). A loading-dependent model of probabilistic cascading failure. *Probability in the Engineering and Informational Sciences.* 19(1), 515-32.

US Department of Energy. *Energy Efficiency Trends in Residential and Commercial Buildings*, available at: `http://http://apps1.eere.energy.gov/buildings/publications/pdfs/corporate/bt_stateindustry.pdf`

US Department of Energy. *Smart Grid System Report*, available at: `http://www.smartgrid.gov/sites/default/files/resources/systems_report.pdf`

Dusabe, D., Munda, J., Jimoh, A.(2009). Modelling of cloudless solar radiation for PV module performance analysis. *Journal of Electrical Engineering* , Vol. 60, NO. 4, 192-197.

Dusko P. N, Dobson I, Daniel S. K., Benjamin A. C. and Vickie E. L.(2006). Criticality in a cascading failure blackout model. *Electrical Power and Energy Systems*, 28 (2006): 627-633

Luo, Q., Ariyur, K. B. and Mathur A. K.(2009), Real Time Energy Management : Cutting the Carbon Footprint and Energy Costs via Hedging, Local Sources and Active Control, *ASME 2009 Dynamic Systems and Control Conference*, Vol. 1, 157-164 .

Machowski, J., Bialek, J and Bumby, B.(2008). *Power System Dynamics: Stability and Control* second edition. Jon Wiley & Sons, Ltd. IBSN: 9780470725580

Murty,PRS.(2008). *Operation and Control in Power Systems*, first edition. BS Publications. IBSN: 9788178001810

North American Electric Reliability Corp.(2009). *Accommodating High Levels of Variable Generation.* Available: `http://www.nerc.com/files/IVGTF_Report_041609.pdf`

Solar Intensity Data is from *National Solar Radiation Data Base*, available at: `http://rredc.nrel.gov/solar/old_data/nsrdb/`

Nuqui, R.(2009). *Electric Power Monitoring with Synchronized Power Measurements*, first edition, VDM Verlag Dr. Muller, ISBN-10: 3639116399

Srivastava, A. and Flueck, A.(2008). *Contingency Screening Techniques And Electric Grid Vulnerabalities*, first edition, VDM verlag, ISBN-10: 3836487012

Singh K.D.P.and Shama S.P.(2009). Enhancement in Thermal Performance of Cylindrical Parabolic Concentrating Solar Collector, *ARISER* Vol. 5 No. 1, 41-48

Temperature data is from *United States Historical Climatology Network*, availabel at: http://cdiac.ornl.gov/epubs/ndp/ushcn/ushcn.html

Section 3

Thermal Application

7

The Summer Thermal Behaviour of "Skin" Materials in Greek Cities as a Decisive Parameter for Their Selection

Flora Bougiatioti
Hellenic Open University
Greece

1. Introduction

The materials, which are used for the pavements of urban spaces and the external renderings of vertical (facades) and horizontal (flat roofs) surfaces of buildings, constitute, the "skin" of a city. In Greece, the selection of these materials is most of the times, based on economic and aesthetic criteria. While the importance of these selection parameters is clear and irrefutable, it should be noted that paving and facade materials play a decisive role on the heat transfer processes, which take place between the city and the climatic environment.

Greece has a warm Mediterranean climate, where the long hours of sunshine and the intensity of solar radiation during the summer result in elevated temperatures of the horizontal and vertical city surfaces in different periods of the day. These high temperatures largely affect the deterioration of the urban heat island (Oke, 1995; Akbari, 1992; Santamouris, 2001a) and thermal comfort conditions in urban open spaces. For the city of Athens, the development of the urban heat island during the summer has grave consequences on the energy consumed for cooling (Santamouris, 2001), and negatively influences many aspects of the citizens' everyday life.

The thermal behaviour of paving materials has been documented by various researchers, both in-situ in urban open spaces (Givoni, 1998; Marques de Almeida, 2002; Cook, et al., 2003; Labaki, et al., 2003) and experimentally on samples building materials (Cook, et al., 2003; Doulos, et al., 2004; Atturo and Fiumi, 2005, Synnefa, et al., 2006). The thermal behaviour of the materials, which form the outer surfaces of building facades, has been investigated in studies. The studies, which were concerned with the effect of shading, and the surface temperatures of various façade materials were carried out by such workers as Hoyano, 1988; Cadima, 1998; Papadakis, et al., 2001; Cantuaria, 2002; Boon Lay, et al., 2000, as well as in a previous study by the author Bougiatioti, et al.(2009).

The study presented in this chapter attempts to combine the two different approaches. It examines the thermal behaviour of paving and façade materials during the summer period, both in-situ in urban open spaces and buildings in Athens, Greece, and samples of building materials placed on a flat roof and exposed to solar radiation during the summer period. The aim of the research is to provide a large set of experimental data (surface temperatures), which can help better in understanding the summer thermal behaviour of the most commonly used surfacing materials in Greece. This experimental data can be qualitatively incorporated as input to the total "image" of each material, in such a way as to be easily evaluated and understood by architects and planners.

The study is divided into two parts: the first part presents the results that concern paving materials and the second presents those involving the facade materials. In each, both the in-situ and the experimental measurements are presented in a cumulative way, in order to draw conclusions on the thermal behaviour of the various building materials during the summer period and the parameters that affect and determine it.

For paving materials, in-situ measurements were conducted in a number of selected urban open spaces in Athens, while experimental measurements involved samples of building materials placed on a flat roof. The materials, which were measured during both the in-situ and the experimental study, were classified into the following general categories: Loose, earthen materials, Natural stone products (slabs of marble, granite), Cement products (gavel concrete, slabs and blocks,), Ceramic products (tiles and blocks), Wooden products (boards), Asphalt products (asphalt concrete), Vegetal surfaces (in-situ measurements) and Water surfaces (in-situ measurements).

For façade materials, in-situ measurements were conducted on the facades of a number of selected buildings, while experimental measurements involved samples of building materials placed on an experimental setting facing, first towards the West and then towards the South. The materials of which temperatures were measured during both the in-situ and the experimental study, were classified into the following general categories: Natural stone products (slabs of marble, granite), Cement products (lime-cement mortar on selected building facades), Ceramic products (tiles and panels), Wooden products (composite panels), Metal products (metal sheets and composite panels), Vegetal surfaces (in-situ measurements of wall covered with climbing plants) and Photovoltaic (PV) panels (in-situ measurements).

The in-situ surface temperatures of paving, as well as of facade materials, are influenced by their contact with the substrate, which, in turn, influences thermal storage and time lag. As a result, the surface temperatures, which are measured on placed materials, can be different from those measured on samples of building materials (Cook, et al., 2003). On the other hand, conducting measurements on samples of building materials exposed to the same environmental conditions provides the ability to simultaneously measure the surface temperatures of a large number of materials. While the measurements on samples of building materials might not reflect their real thermal behaviour, they provide comparative information on the fluctuation of their surface temperatures. Either way, due to the large number of parameters, which influence the surface temperature of building materials in the urban context, it goes without saying that the experimental measurements presented in this chapter should be considered as indicative and not absolute.

The results of the in-situ measurements are presented in Table 1 for paving materials and Table 3 for facade materials, while those of the experimental surface measurements are presented in Table 2 for paving materials and Tables 4 and 5 for facade materials with west and south orientations, respectively.

For all the materials, the differences from the respective air temperature values were calculated in order to obtain a more accurate understanding of the results of the measurements. Furthermore, the air temperature around sunset ($T_{19:30}$) is also mentioned, as it is considered to be indicative of the effect that the materials have on the development of the urban heat island and thermal comfort conditions in the evening.

2. The thermal behaviour of paving materials

2.1.1 Methodology of the in-situ measurements

The choice of the urban open spaces, where the measurements were conducted, is based on a number of criteria, such as: use of commonly applied materials, construction details, reduced overshadowing by adjacent buildings and increased insolation during the day, presence in the same space, of both exposed and shaded areas of the materials and architectural issues, concerning design and patterns of use.

The measurements were conducted in 20 open spaces in Athens, Greece for a month, from June 11th to July 8th. The measurement for each space was carried out for one day. The surface temperature readings were taken with an Optex Thermo-Hunter PT-5LD Infrared (IR) thermometer every half an hour, from 8:00 in the morning until 19:30 or 20:00 in the evening. Only days, which were characterised by predominantly clear skies (0/8) and elevated air temperatures (29.6 ºC to 34.5 ºC), were chosen. Relative humidity values were rather low (28.5 % to 52 %), whereas air velocities ranged from 0.6 m/s to 3.2 m/s.

2.1.2 Methodology of the experimental measurements

During the experimental study, the temperatures of a large number of materials that are usually used for the open spaces in Greek cities, as well as for the flat roofs of buildings, were measured. The experimental measurements were conducted on the flat roof of a building. The temperatures of all the materials were measured for a total period of two weeks. The first day measurements included the night-time readings. The measurements were taken with an Optex Thermo-Hunter PT-5LD Infrared (IR) thermometer from 8:00 to 20:00 at 1-hour intervals. Furthermore, for a one-day period, a contact thermometer (Technoterm 9500) was used simultaneously with the IR one, in order to confirm the accuracy of the IR readings and reveal possible discrepancies. Air temperatures, relative humidity and air velocity were also measured on the site.

In order to define the methodology of the experimental study, a small-scale preliminary study was conducted with samples of cement and ceramic products (slabs and tiles), in order to define the placing mode, which is closest to that of actual conditions (materials placed on a concrete substrate). This was as a result of the fact that the placement of the materials influences their surface temperatures. Three identical cement slabs and three identical ceramic tiles were used with the following placement details: fastened with sand-cement mortar on a dense concrete slab, 30cm thick (base case), set directly on a flat roof constructed with light blue-grey ceramic tiles and laid on a 3cm slab of extruded polystyrene (XPS) painted white.

The results of the preliminary study showed that the samples which were placed on the insulating layer, tended to overheat and develop surface temperatures that were significantly higher than those of the base case. The surface temperatures of the samples, which were directly placed on the flat roof, were higher than those of the base case, but quite close to them. Based on these results, all the samples should have been set on a concrete substrate or placed on gravel or sand substrate. Nevertheless, for the large number (80) of examined materials that the study involved, this was not possible. Consequently, it was decided that the materials would simply be placed on the flat roof of the building, without the interference of another layer.

Fig. 1. Short-scale preliminary study. Different placing modes.

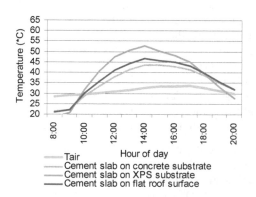

Fig. 2. Short-scale preliminary study. Mean daily temperature fluctuation.

2.2 Results of the in-situ measurements of paving materials

The results of the in-situ measurements, reported here   constitute an overview of the overall study. For this reason, the analysis does not cover every urban space separately, but covers the different categories of materials, in general. The results of the measurements are presented in Table 1.

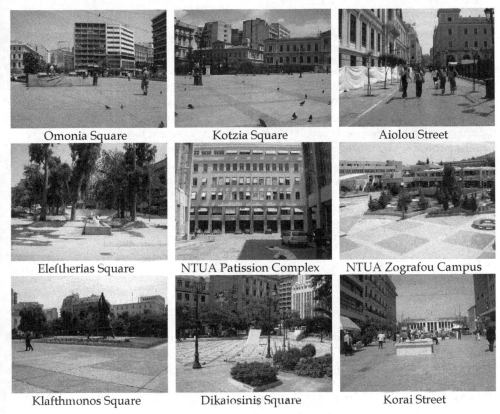

Omonia Square	Kotzia Square	Aiolou Street
Eleftherias Square	NTUA Patission Complex	NTUA Zografou Campus
Klafthmonos Square	Dikaiosinis Square	Korai Street

Fig. 3. Pictures of the some of the urban open spaces, where the in situ measurements were conducted.

2.2.1 Loose, earthen materials

Surfaces of loose materials, such as soil, sand and gravel, are seldom found in the urban open spaces of Athens. The few surfaces that were noted and whose temperatures were measured in this study were completely dry. Until solar noon (around 13:30 local time for Athens, Greece near the summer solstice), the measured surface temperatures were very high (Table 1). After this time, most loose materials surfaces cooled down by 2 to 3 ºC every half hour, reaching the air temperature by sunset (19:30 to 20:00).

2.2.2 Natural stone products

Slabs made of natural stone (mainly granite and marble) are used in urban squares, while stone blocks are used mainly in pedestrian and low-traffic streets. As was expected, white marble slabs were cooler than all the other natural stone materials. But dark grey marble and granite slabs recorded very high maximum surface temperatures, which were over 20 degrees above the corresponding air temperature. At 19:30 these materials were about 8 to 9 ºC hotter than the air. Finally, it was observed that the surface temperatures of the shaded materials were very low.

2.2.3 Cement products

Cement products are the most widely used materials for the open spaces of Athens, and of every other Greek city. Concrete with visible aggregates of various sizes (gravel cement), which is poured on site, is mainly used for the paving of squares, pedestrian streets and urban furniture, such as sitting areas. Slabs are used in sidewalks, while blocks are used in pedestrian and low-traffic streets. As was expected, the light-coloured materials (gravel cement and the white slabs) were cooler than the dark-coloured ones (the grey cement slabs, the red mosaic slabs, the gravel cement slabs and the grey cement blocks). It should be noted that the surface temperatures of all the materials, except of the white slabs, were 3.5 to 5.5 °C hotter than the air around sunset. Also, that the differences in the surface temperatures of exposed and shaded materials were in all the cases significant.

2.2.4 Ceramic products

Ceramic blocks are used in the urban open spaces of Athens in very few cases, mainly in pedestrian streets. In this study, the temperature of the beige ceramic blocks were measured in Korai Pedestrian Street. Due to their colour, the materials did not exhibit very high surface temperatures (Table 1).

2.2.5 Wooden products

Wood is very seldom used in the open spaces of Greek cities. In this study, the temperature of wood surfaces was measured in three cases. Wooden surfaces showed very high surface temperatures, which were comparable even with the surface temperatures of asphalt (Table 1). The mean maximum temperature of wood was 60 °C, its mean temperature was 46.2 °C, while its surface temperature at 19:30 was 28.7 °C, and much lower than the air temperature at that time.

2.2.6 Asphalt products

In the urban open spaces, asphalt products are only used in the form of asphalt concrete, which is widely used for the paving of streets. Asphalt surfaces's temperatures were measured in most urban open spaces. In most cases, their surface temperatures in the afternoon (13:30 - 16:30) were over 50 °C, in some cases, greater than 60 °C. At this point, a distinction should be made between newly and old paving. The weathering of asphalt concrete causes the appearance of its aggregates to be usually white, crushed, light-coloured stone and a subsequent increase in its reflectivity. Apart from the apparent differences in the maximum and mean temperatures, it is interesting to note the surface temperatures of the material around sunset (about 19:30). At this time, air temperatures are around 32 °C, while surface temperatures of new asphalt are 7 °C higher, and that of old asphalt are about 4.5 °C higher, while of shaded asphalt are 2.5 °C lower.

2.2.7 Vegetal surfaces

Vegetal surfaces, such as grass and shrubs, are continuously becoming scarce in the centre of Athens, and their temperatures were measured in few open spaces. As it can be seen in Table 1, in general, vegetal surfaces remain relatively cool during the day. Only the dry grass surface had maximum and mean temperatures equal to 45 and 34.5 °C, while the evergreen shrubs had low maximum and mean temperatures of 34 ºC and 29.5 ºC, respectively. The differences in the surface temperatures of dry and irrigated grass are significant: 10.5 ºC for the maximum temperatures and 6.1 ºC for the mean temperatures. This fact indicates the dependency of the surface temperatures of vegetal surfaces on their irrigation.

2.2.8 Water surfaces

Water surfaces are encountered in the centre of Athens mainly in the form of fountains. The water surfaces, of which temperature were measured in this study were fountains in Kotzia Square and in Eleftherias Square, which showed very low surface temperatures. The obtained maximum and mean surface temperatures were 25.5 and 25.6 ºC, respectively, while their surface temperature around sunset was 25.5 ºC. It was noted that water surfaces, even in the case of non-functioning jet fountains, were constantly cooler than the air.

	T range 13:30 - 16:30 [ºC]	Abs max T [ºC]	Abs min T [ºC]	Mean T [ºC]	Mean T_{air} [ºC]	T at 19:30 [ºC]	T_{air} at 19:30 [ºC]
Loose, Earthen materials							
Earth, dry	49-54.6	54.6	26.4	42.5	32.5	32.8	32.2
Earth, shade	28.3-31.3	31.5	22.0	28.0	32.1	28.3	32.0
Earth-cement mix	45.8-49.4	49.4	24.4	40.4	34.1	35.0	32.0
Earth-cement mix, shade	30-32.5	32.5	26.5	29.1	33.5	31.0	32.0
Gravel	44-47,3	47,3	22,7	37,7	32,5	31,0	32,0
Gravel, shade	31-32,5	32,5	26,0	29,1	32,4	30,0	32,0
Natural stone slabs and blocks							
Marble slabs, white	41.2-44.7	44.7	23.0	36.3	32.6	33.5	31.5
Marble slabs, white, shade	26-31	31.0	20.0	25.0	29.7	25.0	28.0
Marble slabs, dark grey	53.7-58	58	24.7	44.2	33.1	38.7	32.7
Marble slabs, dark grey, shade	29-32	32.0	26.0	28.7	32.4	29.7	32.0
Stone block, grey	50.9-54.4	54.4	26.4	43.9	32.3	37.7	32.0
Stone block, grey, shade	31-34	34.0	25.0	30.1	32.5	29.3	32.0
Stone block, red	48-51	51.0	25.0	42.3	34.5	37.7	32.0
Stone block, red, shade	31-33	33.0	23.0	29.4	34.5	31.0	32.0
Concrete surfaces, Cement slabs and blocks							
Gravel concrete, white	50-52.3	52.3	26.3	41.6	32.2	38.0	32.0
Gravel concrete, white, shade	27-31	31.0	24.0	27.8	32.2	29.0	32.0
Gravel concrete, black	52.3-56	56.0	23.3	43.9	32.2	38.7	32.0
Gravel concrete, black, shade	28-32	32.0	25.0	28.5	32.2	30.0	32.0
Seating areas made of concrete	45.4-48	48.0	25.4	40.1	33.2	35.8	32.2
Cement slabs, white	40.6-47.2	47.2	23.6	37.6	33.5	33.6	32.2
Cement slabs, grey	49-53	53.0	27.3	42.9	33.2	38.3	32.0
Cement slabs, grey, shade	27.3-29	29.0	23.3	26.8	33.2	27.7	32.0
Terrazzo slabs, red	46-51.8	51.8	24.3	41.2	33.6	35.3	32.0
Terrazzo slabs, red, shade	29-31.5	31.5	25.0	29.2	33.8	30.0	32.0
Gravel cement slabs	48.6-52.3	52.3	26.1	42.8	32.6	37.6	31.7
Gravel cement slabs, shade	27.8-30.5	30.5	24.3	27.8	32.8	28.0	31.3
Cement blocks, grey	50.5-53.8	53.8	24.5	43.5	34.1	37.5	32.0
Cement blocks, grey, shade	27.5-29	29.0	23.5	27.0	33.8	28.0	32.0

Ceramic blocks						
Ceramic blocks, brown	46-49	49.0	24.0	37.1	34.5	33.0	32.0
Ceramic blocks, beige, new	43-45	45.0	23.0	35.5	32.2	30.0	32.0
Ceramic blocks, beige, old	46-47	47.0	25.0	39.0	32.2	31.0	32.0
Ceramic blocks, beige, shade	26-29	30.0	24.0	26.7	32.2	28.0	32.0
Wooden surfaces						
Wood (mean)	53.3-59	59.0	29.3	46.2	32.6	28.7	32.3
Asphalt concrete						
Asphalt, old (mean)	48.4-51.4	51.4	27.0	42.1	31.6	36.4	31.7
Asphalt, new (mean)	55.3-59.1	59.1	26.8	46.9	33.7	39.0	32.2
Asphalt, shade (mean)	29.8-32.7	32.7	24.8	28.4	32.7	29.5	32.3
Vegetative cover and Water						
Grass, green, irrigated	31.3-34.5	34.5	19.7	28.8	32.6	24.5	32.0
Grass, dry	41-45	45.0	23.0	34.9	30.8	34.9	32.0
Grass, shade	27-30.5	30.5	18.5	25.0	33.4	25.5	32.0
Bushes, evergreen	29-34	34.0	24.0	29.5	33.8	29.0	32.5
Water (mean)	24.6-26	25.5	24.5	25.6	33.8	25.5	32.0

Table 1. Overview of the in-situ surface temperature measurements in urban open spaces.

2.3 Results of experimental measurements of paving materials

The results of the experimental measurements, are presented in Table 2.

Fig. 4. Experimental measurements of paving materials. Placement of samples on a flat roof surface.

Fig. 5. Experimental measurements of paving materials. Placement of samples on a flat roof surface.

2.3.1 Loose, earthen materials

Loose materials (gravel, light brown sandy soil and dark brown peat soil) were placed in 30x30x4 cm wooden boxes in order to achieve similar dimensions to those of the rest of the materials' samples. The gravel surface was significantly cooler than the other two samples (Table 2). The earth samples' temperature reached mean maximum surface temperatures of between 57.2 °C and 63 °C, depending on their colour. These temperatures were very high (22.7 °C and 28.5 °C higher) compared to the respective mean maximum air temperature. It should be noted that after 15:00 hr, the surface temperatures of the earth samples began to drop steeply, by 4 to 6 °C every hour. As a result, after sunset, the materials were cooler than the air by 2 to 3 °C. The mean temperatures of the earth samples were equal to 42 °C and 45,3 °C and were higher than the mean air temperature by 11.4 °C and 14 °C, respectively.

2.3.2 Natural stone products

The temperature of two different categories of natural stone products were measured: marble and granite slabs and different limestone blocks. As was expected, the white marble slab recorded the lowest surface temperatures (mean maximum 34.8 °C, mean 29.8 °C). Among the marble samples, the dark grey sample was the warmest, measuring maximum temperatures of 53.8 °C, which is 19.3 °C higher than the air temperature, and mean temperatures of 42.9 °C (Tair + 11.6 °C). The surface temperatures of the rest of the marble slabs were between the two above-mentioned extremes (Table 2). Red-black, black and dark grey granite samples recorded higher maximum temperatures than the dark grey marble one, which were 56.4 °C, 56.4 and 58.6 °C, respectively. The mean surface temperatures of these three samples were 43.9 °C, 44.4 °C and 45.1 °C. Finally, the surface temperatures of the various stone blocks, were observed to be mainly dependent on their colours and were similar to those of the corresponding marble samples (Table 2).

2.3.3 Cement products

The concrete samples, which were tested, comprised of white, light-weight concrete (density = 320 kg/m³) and of grey, dense concrete (density = 2400 kg/m³). It is interesting to note (Table 2) that even though the light-weight sample had a lighter colour than the dense concrete one, its surface temperatures were higher (47.1 °C and 12.6 °C higher than the air temperature, compared to 44.4 °C, i.e. Tair+9.9 °C). The mean surface temperatures of the materials were similar and about 36.5 to 37 °C (5 to 6 °C higher than the mean air temperature).

Among the various cement slabs, the coolest was, as expected, the simple, white one. Its mean maximum surface temperature was higher than the air temperature by only 3.3 °C (37.8 °C), while its mean temperature was higher than the mean Tair by only 1 °C (32.3 °C). From the four samples (white, yellow, red and grey), the one with the highest mean maximum temperatures was the red sample (46.5 °C, Tair+12 °C). The striped / textured cement slabs had similar temperature fluctuation with the simple ones (Table 2). This is consistent with the findings of Doulos et al.(2004) who demonstrated that the effect of texture on the surface temperatures of building materials exposed to solar radiation is not statistically important.

The temperatures of Mosaic (terrazzo) slabs with two different grain sizes, and four different colours (white, yellow, red and grey), were measured. The white slabs were obviously the coolest, with mean maximum temperatures of 38.1 °C (Tair+3.6 °C) for the light grain and 39 ℃ (Tair+4.5 ℃) for the heavy grain sample, While   the grey samples had high surface temperatures, reaching 55.6 ℃ (Tair+21.1 ℃) for the light grain sample, and 54.8 ℃ (Tair+20.3 ℃) for the heavy-grain one.

From the pebble cement slabs, the coolest was the one with a surface of white gravel (39.6 ℃, Tair+5.1 ℃), while the warmest was the one with green and grey gravel (53.7 ℃, Tair+19.2 ℃).

The cement blocks used had two different shapes, rectangular and square. The lower limit of the temperature of the white samples and that of the upper limit of the grey-coloured ones determined the range of the surface temperatures of the different colours. White cement blocks reached mean maximum surface temperatures of 41 to 42 ℃ (Tair+7 to 7.5 ℃), while the grey ones were significantly warmer, and their mean maximum temperatures around noon were about 50 ℃ (Tair+15 to 17 ℃).

2.3.4 Ceramic products

This study consisted of four ceramic blocks, two beige and two brown ones, with different thickness (3 and 5 cm). It can be seen (Table 2) that the different thickness of the samples had an insignificant effect (less than 1 ℃ difference) on their mean maximum and mean surface temperatures. The beige samples had mean maximum temperatures of about 43 ℃ around noon, while the brown-coloured ones were warmer, with 48 ℃.

2.3.5 Wooden products

The timber products, which were included in this study, were boards of tropical hardwood (teak and merbau). The merbau sample recorded very high mean maximum surface temperatures, of about 57.4 ℃ (Tair+22.9 ℃) and were comparable to those of the light-coloured, sandy earth sample. The teak sample was relatively cooler with mean maximum temperature of about 52,9 ℃, i.e, Tair+18.4 ℃. Obviously the wood samples had high mean surface temperatures of 40.8 ℃ and 43.9 ℃, for the teak and merbau, respectively.

2.3.6 Asphalt products

The new asphalt concrete sample was black, and for this reason had very high mean maximum temperature of 61.8 ℃, Tair+27.3 ℃ and mean temperature of 46.7 ℃, Tair+15.5 ℃, While the weathered asphalt sample was significantly cooler, having a mean maximum temperature of 50.6 ℃ (Tair+16.1 ℃) and a mean temperature of 40.3 ℃ (Tair+9 ℃). The differences between the new and the weathered samples were larger than 10 ℃, for the mean maximum, and around 7 ℃, for the mean surface temperatures.

Asphalt water-proofing membranes recorded the highest surface temperatures among all the samples. All the samples apart from the one with the polished aluminium facing (mean maximum temperature equal to 46.8 ℃, Tair+12.3 ℃) recorded temperature exceeding the mean maximum surface temperature of 65.5 ℃ (Tair+31 ℃). It should be noted, though, that the samples were placed on an extruded polystyrene slab, which differs from their actual placing practice on a concrete substrate. As a result, the temperatures that were measured may be higher than those, which would have been measured if the materials were placed on a more conductive substrate.

2.3.7 Vegetal surfaces

The vegetal surfaces used for the study were two samples of grass placed on a 4cm thick earth substrate. One of the samples was regularly irrigated, while the other was not. The mean maximum surface temperatures of the samples were very close, about 40 °C (Tair+5.5 °C), while their mean temperatures were lower than the air temperature by less than 1 °C. After sunset (20:00), the samples were significantly cooler (about 5 °C) than the air.

	Mean maximum T [°C]		Mean minimum T [°C]		Mean mean T [°C]	
Air	34.5		28.5		31.3	
Loose, earthen materials						
Gravel	45.5	(+11.0)	22.7	(-5.8)	35.9	(+4.6)
Sandy soil, light brown	57.2	(+22.7)	22.6	(-5.9)	42.7	(+11.4)
Peat soil, dark brown	63	(+28.5)	22.7	(-5.8)	45.3	(+14.0)
Natural stone slabs and blocks						
Marble, 20x30x2 cm, white	34.8	(+0.3)	21.8	(-6.7)	29.8	(-1.5)
Marble, 20x30x2 cm, light grey	41.4	(+6.9)	22	(-6.5)	34	(+2.7)
Marble, 20x30x2 cm, beige	44.8	(+10.3)	23.4	(-5.1)	36.6	(+5.3)
Marble, 20x30x2 cm, grey-beige	49.4	(+14.9)	23.7	(-4.8)	39.4	(+8.1)
Marble, 20x30x2 cm, grey	49.4	(+14.9)	24	(-4.5)	40	(+8.7)
Marble, 20x30x2 cm, ochre	50	(+15.5)	23.4	(-5.1)	39.8	(+8.5)
Marble, 20x30x2 cm, dark red	52.4	(+17.9)	23.8	(-4.7)	40.6	(+9.3)
Marble, 20x30x2 cm, dark grey	53.8	(+19.3)	23.9	(-4.6)	42.9	(+11.6)
Granite, 20x30x2 cm, white-beige	47.6	(+13.1)	26.3	(-2.2)	38.8	(+7.5)
Granite, 20 x 30 x 2 cm, salmon	48.2	(+13.7)	25.9	(-2.6)	39.2	(+7.9)
Granite, 20 x 30 x 2 cm, red-black	56.4	(+21.9)	25.9	(-2.6)	43.9	(+12.6)
Granite, 20 x 30 x 2 cm, black	56.4	(+21.9)	26.2	(-2.3)	44.4	(+13.1)
Granite, 20 x 30 x 2 cm, dark grey	58.6	(+24.1)	27.4	(-1.1)	45.1	(+13.8)
Blocks, 10 x 10 x 3 cm, white	42.5	(+8)	22.5	(-6)	35	(+3.7)
Blocks, 10 x 10 x 3 cm, white-black	45.9	(+11.4)	20.3	(-8.2)	36.2	(+4.9)
Blocks, 10 x 10 x 3 cm, grey-red	46.9	(+12.4)	21.2	(-7.3)	37.3	(+6)
Blocks, 10 x 10 x 3 cm, red	47.9	(+13.4)	21.4	(-7.1)	38	(+6.7)
Blocks, 10 x 10 x 3 cm, grey	48.3	(+13.8)	21	(-7.5)	37.7	(+6.4)
Blocks, 10 x 10 x 3 cm, red	49.2	(+14.7)	21.4	(-7.1)	38.4	(+7.1)
Blocks, 20 x 10 x 5 cm, beige	45.3	(+10.8)	22.2	(-6.3)	36.3	(+5)
Blocks, 20 x 10 x 5 cm, dark red	48.4	(+13.9)	21.6	(-6.9)	38	(+6.7)
Blocks, 20 x 10 x 5 cm, grey	51.3	(+16.8)	21.9	(-6.6)	39.9	(+8.6)
Concrete surfaces, Cement slabs and blocks						
Light-weight concrete, white	47.1	(+12.6)	22.2	(-6.3)	37.1	(+5.8)
Dense concrete, (30x30x9 cm), grey	44.4	(+9.9)	21.7	(-6.8)	36.4	(+5.1)
Slabs, 30x30x3 cm, simple, white	37.8	(+3.3)	22.6	(-5.9)	32.3	(+1)
Slabs, 30x30x3 cm, simple, yellow	45	(+10.5)	23.2	(-5.3)	37	(+5.7)
Slabs, 30x30x3 cm, simple, grey	45.6	(+11.1)	23.3	(-5.2)	37.4	(+6.1)
Slabs, 30x30x3 cm, simple, red	46.5	(+12)	24.5	(-4)	38.2	(+6.9)
Slabs, 30x30x3 cm, striped, yellow	44	(+9.5)	22.6	(-5.9)	36.1	(+4.8)
Slabs, 30x30x3 cm, striped, red	46.7	(+12.2)	22.9	(-5.6)	37.9	(+6.6)
Slabs, 30x30x3 cm, striped, grey	48	(+13.5)	23.1	(-5.4)	38.9	(+7.6)
Slabs, 30x30x3 cm, mosaic, white	38.1	(+3.6)	22.1	(-6.4)	32.5	(+1.2)
Slabs, 30x30x3 cm, mosaic, yellow	42	(+7.5)	22.5	(-6.0)	35	(+3.7)
Slabs, 30x30x3 cm, mosaic, red	52.5	(+18)	23.3	(-5.2)	42	(+10.7)
Slabs, 30x30x3 cm, mosaic, grey	55.6	(+21.1)	23.2	(-5.3)	43.7	(+12.4)

Slabs, 30x30x3 cm, mosaic, white	39	(+4.5)	22.1	(-6.4)	33	(+1.7)
Slabs, 30x30x3 cm, mosaic, yellow	42.6	(+8.1)	22.4	(-6.1)	35.4	(+4.1)
Slabs, 30x30x3 cm, mosaic, red	50.4	(+15.9)	23.1	(-5.4)	40.3	(+9)
Slabs, 30x30x3 cm, mosaic, grey	54.8	(+20.3)	23.3	(-5.2)	43.2	(+11.9)
Slabs, 40x40x3,5 cm, mosaic, white	39.6	(+5.1)	21.6	(-6.9)	32.6	(+1.3)
Slabs, 40x40x3,5 cm, mosaic, yellow	43	(+8.5)	21.4	(-7.1)	34.8	(+3.5)
Slabs, 40x40x3,5 cm, mosaic, red	46.7	(+12.2)	21.8	(-6.7)	37.1	(+5.8)
Slabs, 40x40x3,5 cm, mosaic, grey	52.2	(+17.7)	22.1	(-6.4)	40.8	(+9.5)
Slabs, 40x40x3,5 cm, pebbles, white	39.6	(+5.1)	21.7	(-6.8)	32.6	(+1.3)
Slabs, 40x40x3,5, pebbles, various	50.2	(+15.7)	22.1	(-6.4)	39.4	(+8.1)
Slabs, 40x40x3,5cm, pebbles, grey	51.3	(+16.8)	22.2	(-6.3)	40	(+8.7)
Slabs, 40x40x3,5cm, pebbles, grey	53.4	(+18.9)	22.8	(-5.7)	41.4	(+10.1)
Slabs, 40x40x3,5cm, pebbles, green	53.7	(+19.2)	22.8	(-5.7)	41.8	(+10.5)
Blocks, 20x10x6 cm, white	42	(+7.5)	22.3	(-6.2)	34.7	(+3.4)
Blocks, 20x10x6 cm, yellow	43	(+8.5)	22.8	(-5.7)	35.6	(+4.3)
Blocks, 20x10x6 cm, red	47.7	(+13.2)	23	(-5.5)	38.8	(+7.5)
Blocks, 20x10x6 cm, grey	51.7	(+17.2)	23.1	(-5.4)	41.3	(+10)
Blocks, 10x10x6 cm, white	41	(+6.5)	22.3	(-6.2)	33.9	(+2.6)
Blocks, 10x10x6 cm, yellow	41.8	(+7.3)	22.8	(-5.7)	34.7	(+3.4)
Blocks, 10x10x6 cm, ochre	44	(+9.5)	23.1	(-5.4)	36.2	(+4.9)
Blocks, 10x10x6 cm, beige	44.6	(+10.1)	22.8	(-5.7)	36.6	(+5.3)
Blocks, 10x10x6 cm, light brown	44.7	(+10.2)	23.1	(-5.4)	36.8	(+5.5)
Blocks, 10x10x6 cm, red	46.3	(+11.8)	23.5	(-5)	38.1	(+6.8)
Blocks, 10x10x6 cm, grey	46.7	(+12.2)	23.3	(-5.2)	38.2	(+6.9)
Blocks, 10x10x6 cm, dark brown	49.6	(+15.1)	23.5	(-5)	40.2	(+8.9)
Blocks, 10x10x6 cm, dark grey	49.6	(+15.1)	23.6	(-4.9)	40.3	(+9)
Ceramic blocks						
Beige, 3 cm-thick	42.7	(+8.2)	21.8	(-6.7)	35.2	(+3.9)
Beige, 5 cm-thick	43.6	(+9.1)	21.9	(-6.6)	35.7	(+4.4)
Brown, 5 cm-thick	47.5	(+13.0)	22.4	(-6.1)	37.9	(+6.6)
Brown, 3 cm-thick	48.3	(+13.8)	22.8	(-5.7)	38.7	(+7.4)
Firebrick, ochre	45.1	(+10.6)	22.7	(-5.8)	36.3	(+5.0)
Firebrick,, brown	48.4	(+13.9)	23	(-5.5)	38.2	(+6.9)
Wooden surfaces						
Teak	52.9	(+18.4)	22.7	(-5.8)	40.8	(+9.5)
Merbau	57.4	(+22.9)	23.3	(-5.2)	43.9	(+12.6)
Asphalt						
Asphalt concrete, old, grey	50.6	(+16.1)	22.5	(-6)	40.3	(+9)
Asphalt concrete, new, black	61.7	(+27.2)	23.4	(-5.1)	46.4	(+15.1)
Asphalt concrete, new, black	61.9	(+27.4)	23	(-5.5)	47.1	(+15.8)
Asphalt membrane, alu cover	46.8	(+12.3)	28.1	(-0.4)	38.7	(+7.4)
Asphalt membrane, black	67.2	(+32.7)	21.3	(-7.2)	47	(+15.7)
Asphalt membrane, grey light-grain	70	(+35.5)	21.8	(-6.7)	48.7	(+17.4)
Asph. membrane, grey heavy-grain	65.5	(+31)	20.7	(-7.8)	45.9	(+14.6)
Asphalt membrane, black	72.7	(+38.2)	22.7	(-5.8)	50.3	(+19)
Vegetative cover						
Grass, well-irrigated	39.9	(+5.4)	20	(-8.5)	31.1	(-0.2)
Grass, not irrigated	40.4	(+5.9)	20.1	(-8.4)	30.9	(-0.4)

Table 2. Overview of the experimental surface temperature measurements on samples of paving materials.

3. The thermal behaviour of facade materials

3.1.1 Methodology of the in-situ measurements

The choice of the buildings selected for the study was based on a number of criteria, which are summarised below:

- Buildings with facades oriented towards the four basic orientations (north, south, east and west).
- Free-standing buildings with more than one facades which is -as far as possible-from obstruction of solar access, which would allow the comparison of the surface temperatures recorded at the different orientations.
- Buildings with western orientation, and are most thermally stressed   during the summer months.
- Buildings with facades constructed of contemporary materials (e.g. panels of granite, composite aluminium panels, etc.) mainly with the ventilated facade system.
- Groups of buildings, which would allow the comparison of the surface temperatures of the different materials with the same orientation.

The surface temperature readings were taken with an Optex Thermo-Hunter PT-5LD Infrared (IR) thermometer every half-hour, from 8:00 in the morning until 19:30 or 20:00 in the evening. The measurements were limited to a height of not more than 6 meters, due to restrictions posed by the instrument. On each of the facades, IR readings were taken at different heights and at different parts of the facade, in order to ensure that recorded surface temperature was representative. Finally, air temperatures, relative humidity and air velocity were also measured on the site. Indoor air and surface temperature measurements were not possible, as there was no access to the buildings in question. The environmental conditions during the days of the measurements were characterised predominantly by clear skies (0/8), high air temperatures (29.6 °C to 34.4 °C), low relative humidity (28.5 % to 52 %) and various air velocities (0.6 m/s to 3.2 m/s).

3.1.2 Methodology of the experimental measurements

The materials, which temperatures were measured during the experimental study, were mainly materials that are used for building facades with ventilated facade system. Forty-four (44) samples were selected and mounted on two experimental arrangements, which aimed at simulating the ventilated facade system. Each experimental arrangement comprised a 30mm insulating board (extruded polystyrene - XPS), which was fixed on an 18mm chipboard. The samples for surfacing materials were mounted with the use of metal screws, which suspended the materials at a distance (air gap) of 20 to 30 mm from the insulation layer. The distance between the samples was to ensure that there was no contact between them, while each screw was individually "planted" into the insulating to avoid thermal bridges. In all the temperatures of seven (7) samples of marble slabs and five (5) samples of granite slabs were measured, making a total of 56 samples.

The measurements took two weeks: one week (7 days) for the western orientation (10-17/07/2004) and the other week for the southern orientation (31/07-06/08/2004). The measurements were by an Optex Thermo-Hunter PT-5LD Infrared (IR) thermometer, from

8:00 to 20:00 at 1-hour intervals. One day of the measurements included night-time readings (from 20:00 to 08:00) in order to examine the cooling rate of the materials. Furthermore, for a one-day period, a contact thermometer (Technoterm 9500) was used simultaneously with the IR one, in order to confirm the accuracy of the IR readings and account for possible discrepancies. Air temperatures, relative humidity and air velocity measurements were also done at the site.

3.2 Results of the in-situ measurements of facade materials

The results are presented in Table 3.

3.2.1 Natural stone products

Light-coloured stone panels, such as white marble and beige sandstone did not record high surface temperatures in any of the four basic orientations (north, south, west and east). Furthermore, for all the examined orientations, their temperature at $T_{19:30}$ was higher than the corresponding air temperature by only 1 to 2 °C. However, the panels of dark-coloured stone, namely grey and black granite, tended to overheat, with surface temperatures that may well exceed 50 °C for facades which face towards the east and the west. It is also important to note that for west-facing facades, clad with grey or black granite, the surface temperatures around sunset ($T_{19:30}$) were a lot higher than the air temperature (15 °C and 17 °C, respectively).

3.2.2 Cement products

The west-facing reinforced concrete surface, recorded a surface temperature of 47 °C, and was, at 19:30 warmer than the air by 10 °C.

The building facades of typical construction (brick wall with external rendering of lime-cement mortar painted in different colours), were of the following colours: white, beige, grey and dark red. As was expected, the white-painted surfaces remained relatively cool, with maximum temperatures that did not exceed 40 °C and mean temperature that was similar to the mean air temperature. The beige surfaces with western orientation had relatively low mean temperatures, maximum temperatures around 45 °C and were warmer than the air by 4 °C, around sunset. The highest temperatures on western facades were measured by the dark red and grey surfaces. Their maximum temperatures were 56 °C for the dark red facade and 60 °C for the grey one, while their respective $T_{19:30}$ temperatures were higher than the air temperature by 6 °C and 12 °C. Finally, these surfaces had observed mean temperatures, which were higher than the respective air temperature by 3 °C and 7.5 °C.

3.2.3 Ceramic products

The in-situ study was a building, whose four facades were covered with dark brown ceramic blocks. Based on the IR readings, the eastern facade reached a maximum temperature of 38 °C, and the western one, a temperature of 44 °C. Around sunset, the ceramic blocks of the western facades were warmer than the air by 6 °C.

Material	Orientation	Abs max T [°C]	Abs. min T [°C]	Mean T [°C]	Mean T_{air} [°C]	T at 19:30 [°C]	T_{air} at 19:30 [°C]
Marble slabs							
Marble, white	N	36.0	21.0	28.3	29.7	29.0	28.0
Marble, white	S	33.0	20.0	28.0	29.7	25.0	28.0
Marble, white	W	38.0	18.0	27.4	29.7	30.0	28.0
Sandstone slabs							
Sandstone	N	35.8	25.5	30.7	32.6	31.5	31.5
Sandstone	S	39.5	26.0	33.9	33.2	34.0	33.0
Sandstone	E	39.5	25.5	34.5	33.2	34.0	33.0
Sandstone	W	41.6	23.6	30.8	31.8	32.2	31.0
Granite slabs							
Granite, black	N	42.0	21.0	30.7	29.7	34.0	28.0
Granite, black	S	43.5	26.5	36.6	32.1	31.0	30.0
Granite, black	E	53.0	32.5	40.7	34.4	32.5	32.0
Granite, black	W	57	23.0	36.9	32.1	46.5	30.0
Granite, grey	N	41.0	30.0	36.1	34.4	33.5	32.0
Granite, grey	S	52.0	27.0	40.5	34.4	31.5	32.0
Granite, grey	E	49.5	31.5	39.5	34.4	31.5	32.0
Granite, grey	W	62.0	25.0	42.6	34.4	47	32.0
Reinforced concrete							
Reinforced concrete	N	31.0	27.0	29.0	30.8	30.0	32.0
Reinforced concrete	S	38.0	22.0	30.3	30.8	31.0	32.0
Reinforced concrete	E	36.0	29.0	33.3	30.8	30.0	32.0
Reinforced concrete	W	47.0	20.0	33.1	30.8	42.0	32.0
Standard Greek construction (Lime-cement mortar and painting with different colours)							
Standard construction, white	S	33.0	25.7	29.8	32.8	32.0	30.7
Standard construction, white	E	35.0	29.0	32.2	34.4	30.0	32.0
Standard construction, white	W	40.0	20.0	29.5	29.7	28.0	28.0
Standard construction, beige	N	34.0	25.5	31.1	33.2	32.0	33.0
Standard construction, beige	S	38.3	25.3	32.9	32.7	32.6	32.3
Standard construction, beige	E	42.0	25.8	35.8	33.2	33.3	33.0
Standard construction, beige	W	45.3	25.0	33.3	33.4	36.7	32.9
Standard construction, grey	S	44.0	29.0	36.8	34.4	32.0	32.0
Standard construction, grey	W	60.0	27.0	41.9	34.4	44.0	32.0
Standard construction, dark grey	N	36.0	26.0	32.2	33.2	33.0	33.0
Standard construction, dark grey	S	41.0	25.0	34.8	33.2	36.0	33.0
Standard construction, dark grey	W	56.0	28.0	36.2	33.2	39.0	33.0
Composite aluminium panels							
Composite alu panel, dark grey	S	40.0	26.0	32.8	29.7	25.0	28.0
Composite alu panel, dark grey	E	49.5	23.5	32.5	29.7	25.0	28.0
Composite alu panel, dark grey	W	45.0	21.0	31.7	29.7	31.0	28.0
Ceramic bricks (veneer)							
Ceramic bricks, brown	N	36.0	22.0	28.2	29.7	32.0	28.0
Ceramic bricks, brown	S	36.0	22.0	30.7	29.7	27.5	28.0
Ceramic bricks, brown	E	38.0	23.0	30.8	29.7	27.0	28.0
Ceramic bricks, brown	W	44.0	21.0	31.9	29.7	34.0	28.0
PV panels and Climbing plants							
PV panels (hybrid facade)	S	39.2	24.4	31.5	34.0	28.8	33.0
Wall with climbing plants	S	34.0	26.0	30.7	34.0	30.0	33.0

Table 3. Overview of the in-situ measurements on building facades.

3.2.4 Metal products

The metal products involved were grey-coloured composite aluminium panels. In the case of east-facing panels, the maximum surface temperature recorded was 49.5 °C, but for the west-facing ones, it was 45 °C. The temperature, $T_{19:30}$ of the panels was higher than the respective air temperature by 3 °C.

3.2.5 Vegetal surfaces

The vegetal surface was a reinforced concrete wall covered with Virginia creeper (Parthenocissus spp). Its temperatures remained lower than the air temperatures throughout the day.

3.2.6 Photovoltaic (PV) panels

The PV panels used were blue-coloured poly-crystalline ones. These were integrated into a south-facing, ventilated PV façade. Their mean surface temperature was lower than the mean air temperature, while their absolute maximum temperature reached 44 °C. It is important to note that, around sunset, the PV panels were cooler than the air by 4.2 °C. These findings are considered very important because they show that during the summer period, southern facades clad with PV panels do not affect thermally the microclimate, while contributing to the use of renewable energy.

3.3 Results of the experimental measurements of facade materials

The results of the experimental measurements are presented in Table 4 for the western orientation and in Table 5 for the southern orientation.

3.3.1 Western orientation

The environmental conditions throughout the week of measurements were characterised by predominantly clear skies (0/8), high air temperatures (mean maximum air temperature of 34.5 °C, mean minimum air temperature of 28.5 °C and mean air temperature of 31.3 °C).

3.3.1.a Natural stone products

When facing towards the west, marble panels recorded mean maximum surface temperatures of 39.6 °C (Tair + 5.1 °C) for white marble and 58.1 °C (Tair + 23.6 °C) for dark grey marble. The mean surface temperatures of these materials were 31.6 °C (Tair + 0.3 °C) and 39.8 °C (Tair + 8.5 °C), respectively. It should be noted that at 20:00 (after sunset) the surface temperatures of all the samples except of the white one were over 35 °C, indicating that their temperatures were higher than the corresponding air temperature, by more than 5 °C.

All the west-facing granite panels, overheated during the afternoon, reaching mean maximum temperatures of 48.1°C that were well above 45 °C for the white-beige sample, 59.1 °C for the black one. The mean surface temperatures of the samples range from 35.5 °C

(Tair + 4.2 ᵒC) to 40.3 ᵒC (Tair + 9 ᵒC). Similar to the marble samples, the granite samples were, at 20:00, warmer than the air by 5 ᵒC to 10 ᵒC. It is noted that the dark-coloured granite surfaces, were warmer than the air by more than 15 ᵒC.

3.3.1.b Ceramic products

Thin ceramic panels and tiles that face towards the west recorded maximum surface temperatures, which may range from 44 ᵒC (for white) to 56.6 ᵒC (for black) and are from 9.5 ᵒC to 22.1 ᵒC higher than the mean maximum air temperature. The mean surface temperature of the white sample is 33.6 ᵒC (Tair + 2.3 ᵒC) and of the black sample 38.8 ᵒC (Tair + 7.5 ᵒC). The mean surface temperatures of the rest of the samples are between the above-mentioned limits. Finally, the surface temperatures of all the samples of thin ceramic tiles and panels after the sunset (at 20:00) were between 30 ᵒC and 35 ᵒC, i.e., no more than 5 ᵒC higher than the corresponding air temperature.

Ceramic panels of increased thickness (13 mm) recorded surface temperatures similar to the thin ceramic tiles and panels with similar colours. The white-coloured sample has a mean maximum temperature of 44.1 ᵒC (Tair + 9.6 ᵒC) and a mean temperature of 33.8 ᵒC (Tair + 2.5 ᵒC), while the brown-coloured one has mean maximum temperature of 52 ᵒC (Tair + 17.5 ᵒC) and a mean temperature of 37.4 ᵒC (Tair + 6.1 ᵒC).

3.3.1.c Wooden products

West-facing composite wood panels recorded very high surface temperatures. The light-coloured and the intermediate samples reached a maximum temperature of about 52 ᵒC, while the dark-coloured one had a mean maximum temperature of 59.6 ᵒC, which was 25 ᵒC higher than the mean maximum air temperature. Concerning the mean surface temperatures, the samples were warmer than the air by 6 ᵒC to 9.4 ᵒC. After the sunset, the surface temperatures of all three samples were close to the air temperature.

	Mean maximum T [ᵒC]		Mean minimum T [ᵒC]		Mean mean T [ᵒC]	
Air	34.5		28.5		31.3	
Composite aluminium panels 20 x 30 cm facing West						
White colour	44.5	(+10)	24.5	(-4)	33.7	(+2.4)
Yellow colour	48.7	(+14.2)	25.9	(-2.6)	36.3	(+5)
Red colour	51	(+16.5)	25.7	(-2.8)	37.0	(+5.7)
Silver colour	52.9	(+18.4)	25.2	(-3.3)	37.4	(+6.1)
Green colour	55.6	(+21.1)	26.2	(-2.3)	39.1	(+7.8)
Blue colour	56.9	(+22.4)	25.7	(-2.8)	39.3	(+8)
Black colour	66.3	(+31.8)	26.1	(-2.4)	42.4	(+11.1)
Composite wood panels 20 x 30 cm facing West						
Light-coloured	51.8	(+17.3)	26	(-2.5)	37.3	(+6)
Intermediate	52.5	(+18)	25.9	(-2.6)	37.8	(+6.5)
Dark-coloured	59.6	(+25.1)	26.2	(-2.3)	40.7	(+9.4)
Metal sheets 10 x 10 cm facing West						
Copper, natural colour	41.6	(+7.1)	27	(-1.5)	34.0	(+2.7)
Copper, verdigris	48.2	(+13.7)	25.7	(-2.8)	35.7	(+4.4)
Titanium	40.1	(+5.6)	25.4	(-3.1)	32.5	(+1.2)
Zinc	41	(+6.5)	25.4	(-3.1)	31.9	(+0.6)

Quartz zinc, beige	47.1	(+12.6)	26.2	(-2.3)	35.7	(+4.4)
Quartz zinc, turquoise	48.2	(+13.7)	26.1	(-2.4)	36.2	(+4.9)
Quartz zinc, light blue	48.8	(+14.3)	26.1	(-2.4)	36.5	(+5.2)
Quartz zinc, natural, light grey	44.7	(+10.2)	25.7	(-2.8)	35.0	(+3.7)
Quartz zinc, natural, dark grey	49.5	(+15)	26	(-2.5)	36.8	(+5.5)
Ceramic tiles / panels 10 x 10 cm facing West						
White colour (glazed)	44	(+9.5)	25.1	(-3.4)	33.6	(+2.3)
Ochre colour	48.4	(+13.9)	25.1	(-3.4)	35.9	(+4.6)
Beige colour	48.8	(+14.3)	25.5	(-3)	36.4	(+5.1)
Light grey colour	49.8	(+15.3)	25.1	(-3.4)	36.4	(+5.1)
Brown colour	51.9	(+17.4)	25.8	(-2.7)	38.1	(+6.8)
Red-Dark red colour, black spots	52.4	(+17.9)	25.6	(-2.9)	37.1	(+5.8)
Olive green colour, black spots	52.5	(+18)	25.3	(-3.2)	37.2	(+5.9)
Dark grey colour	52.8	(+18.3)	26.2	(-2.3)	38.0	(+6.7)
Blue colour	53.1	(+18.6)	25.9	(-2.6)	38.2	(+6.9)
Black colour	54.3	(+19.8)	26.2	(-2.3)	39.0	(+7.5)
Black colour, with white spots	56.6	(+22.1)	25.8	(-2.7)	38.8	(+7.5)
Ceramic panels 25 x 13 x 1,3 cm facing West						
White colour	44.1	(+9.6)	25.9	(-2.6)	33.8	(+2.5)
Beige colour	46.4	(+11.9)	25.9	(-2.6)	35.3	(+4)
Light grey colour	47.2	(+12.7)	25.7	(-2.8)	35.2	(+3.9)
Brown colour	52	(+17.5)	25.2	(-3.3)	37.4	(+6.1)
Marble panels 20 x 30 cm facing West						
White colour	39.6	(+5.1)	24.2	(-4.3)	31.6	(+0.3)
Light grey colour	46	(+11.5)	24.5	(-4)	34.2	(+2.9)
Beige colour	49.1	(+14.6)	24.6	(-3.9)	35.5	(+4.2)
Grey-Beige colour	51.2	(+16.7)	24.6	(-3.9)	36.5	(+5.2)
Grey colour	54.4	(+19.9)	24.5	(-4)	37.7	(+6.4)
Ochre colour	55.4	(+20.9)	24.9	(-3.6)	38.1	(+6.8)
Dark grey colour	58.1	(+23.6)	25	(-3.5)	39.8	(+8.5)
Granite panels 20 x 30 cm facing West						
White-Beige colour	48.1	(+13.6)	24.7	(-3.8)	35.5	(+4.2)
Salmon colour	47.7	(+13.2)	25.6	(-2.9)	35.3	(+4)
Red-Black colour	56.6	(+22.1)	26.1	(-2.4)	39.4	(+8.1)
Black colour	57.5	(+23)	25.7	(-2.8)	39.8	(+8.5)
Grey colour	59.1	(+24.6)	26	(-2.5)	40.3	(+9)

Table 4. Overview of the experimental measurements on samples of building materials facing West.

3.3.1.d Metal products

When facing towards the west, even light-coloured composite aluminium panels (white colour) recorded surface temperatures, which were higher than the corresponding air temperature by more than 10 ºC. For the range of colours, mean maximum temperatures were 10 ºC (for white colour) to 31.8 ºC (for black colour) higher than the mean maximum air temperature. Mean surface temperatures range from 33.7 ºC for white to 42.4 ºC for black panels, and are thus 2.4 ºC to 11.1 ºC higher than the mean air temperature. Finally, the surface temperatures at 20:00 (after the sunset) are between 30 ºC and 35 ºC and are no more than 5 ºC higher than the air temperature at that time.

The mean maximum surface temperatures of west-facing samples of different metal sheets had a temperature range from 40.1 ºC (titanium) to 49.5 ºC (quartz zinc, dark grey natural). Their mean surface temperatures ranged from 31.9 ºC (polished zinc) to 36.8 ºC (quartz zinc, dark grey natural), and were higher than the respective air temperature by 0.6 ºC to 5.5 ºC.

3.3.2 Southern orientation

The atmospheric conditions throughout the week of measurements were clear skies (0/8), high air temperatures (mean maximum air temperature of 34.7 ºC, mean minimum air temperature of 28.3 ºC and mean air temperature of 31.2 ºC).

3.3.2.a Natural stone products

South-facing marble panels measured lower surface temperatures than west-facing ones. The mean maximum temperature of the white sample was equal to 36.7 ºC (Tair + 2 ºC), which is 3 ºC lower than respective temperature of the west-facing one. Similarly, the mean maximum temperature of the dark grey sample was 51.7 ºC (Tair + 17 ºC), and 6.6 ºC lower than the temperature of the one facing west. The mean surface temperatures of both samples were almost equal to those measured for the western orientation, namely 31.1 ºC (Tair - 0.1 ºC) and 39.7 ºC (Tair + 8.5 ºC), respectively. The surface temperatures of the rest of the samples fall between the above-mentioned extremes.

When granite panels have a southern orientation (Figure 23), their surface temperatures are lower compared to those measured when they faced towards the west. The mean maximum temperatures of all the five south facing samples were 6 ºC to 9 ºC lower than the "western" ones. The white-beige sample had a mean maximum temperature of 39.3 ºC (Tair + 4.6 ºC), while the grey-coloured samples had 50.3 ºC (Tair + 15.6 ºC). Contrary to the marble and the granite samples with southern orientation had lower mean surface temperatures than those of western orientation. The mean surface temperatures of the samples range from 32.8 ºC (Tair + 1.6 ºC) to 38.3 ºC (Tair + 7.1 ºC). The marble and the granite samples, when placed facing south, gradually cool down during the afternoon. As a result, their surface temperatures after the sunset (20:00) are close to the corresponding air temperature, surpassing it by only 1 ºC to 2 ºC.

3.3.2.b Ceramic products

Thin ceramic panels and tiles that face towards the south recorded maximum surface temperatures, which may range from 38.1 ºC (for white) to 50.5 ºC (for black) and are from 3.4 ºC to 15.8 ºC higher than the mean maximum air temperature. The mean maximum temperatures of all the samples were approximately 6 ºC to 7 ºC lower than the mean maximum values, which were measured when the materials had a western orientation. The mean surface temperature of the white sample is 31.6 ºC (Tair + 0.4 ºC) and of the black sample 37 ºC (Tair + 5.8 ºC). The mean surface temperatures of the rest of the samples lie between the above-mentioned limits. Finally, the surface temperatures of all the samples of thin ceramic tiles and panels after the sunset (at 20:00) were close to the corresponding air temperature.

South-facing ceramic panels of thickness 13 mm recorded similar surface temperatures to that of the thin ceramic tiles and panels with similar colours. The white-coloured sample had a mean maximum temperature of 39.7 °C (Tair + 5 °C) and a mean temperature of 32.6 °C (Tair + 1.4 °C), while the brown-coloured one has mean maximum temperature of 48.9 °C (Tair + 14.2 °C) and a mean temperature of 37 °C (Tair + 5.8 °C). Compared to the temperatures, which were measured when the samples were facing west, the above-mentioned surface temperatures are considerably lower.

3.3.2.c Wooden products

The surface temperatures of the composite wood panels with southern orientation were not very high. The light-coloured and the intermediate sample reached a maximum temperature of about 42.3 °C, while the dark-coloured one had a mean maximum temperature of 47.6 °C, which was 12.9 °C higher than the mean maximum air temperature. Concerning the mean surface temperatures, the samples were warmer than the air by 2.6 °C to 5.1 °C. A quick comparison with the temperatures measured for the western orientation reveals that the mean maximum temperature of the light-coloured sample is lower by about 10 °C, while that of the dark-coloured one by about 12 °C. Concerning the mean surface temperatures, the differences between the "western" and the "southern" surface temperatures range between 2.4 °C for the light-coloured sample and 4.3 °C for the dark-coloured one. Again, the observed differences due to the change of orientation are significant.

3.3.2.d Metal products

When facing towards the south, light-coloured composite aluminium panels (white colour) recorded low surface temperatures, but higher than the corresponding air temperature by only 2.6 °C. For the range of colours, mean maximum temperatures ranged from 2.6 °C (for white colour) to 16.5 °C (for black colour) higher than the mean maximum air temperature. The differences with the measured mean maximum temperatures of the western orientation are impressive and range from 7.4 °C, for the white colour, to 15.3 °C, for black colour. Mean surface temperatures ranged from 31.5 °C for white to 38.7 °C for black panels, and are thus 0 °C to 7 °C higher than the mean air temperature. The differences between the "western" and the "southern" mean surface temperatures of the two extreme colours (white and black) are 2.7 °C and 4.1 °C, respectively. Finally, the surface temperatures at 20:00 (after the sunset) are below 5 °C than that of the air temperature at that time.

	Mean maximum T [°C]		Mean minimum T [°C]		Mean mean T [°C]	
Air	34.7		28.3		31.2	
Composite aluminium panels 20 x 30 cm facing South						
White colour	37.3	(+2.6)	22.5	(-5.8)	31.5	(-0.3)
Yellow colour	41.6	(+6.9)	23.8	(-4.5)	33.6	(+2.4)
Red colour	42.5	(+7.8)	24.3	(-4)	34.4	(+3.2)
Silver colour	44.3	(+9.6)	23.6	(-4.7)	34.8	(+3.6)
Green colour	46.4	(+11.7)	24.5	(-3.8)	35.8	(+4.6)
Blue colour	47.2	(+12.5)	24.5	(-3.8)	36.8	(+5.6)
Black colour	51.2	(+16.5)	25.1	(-3.2)	38.2	(+7)
Composite wooden panels 20 x 30 cm facing South						
Light-coloured	42.3	(+7.6)	24.1	(-4.2)	33.8	(+2.6)
Intermediate	43.3	(+8.6)	24.2	(-4.1)	34.1	(+2.9)
Dark-coloured	47.6	(+12.9)	24.8	(-3.5)	36.3	(+5.1)

Metal sheets 10 x 10 cm facing South					
Copper, natural colour	41.2	(+6.5)	26.6	(-1.7)	34.7	(+3.5)
Copper, verdigris	44.1	(+9.4)	24.8	(-3.5)	34.4	(+3.2)
Titanium	39.5	(+4.8)	26.1	(-2.2)	33.1	(+1.9)
Zinc	40.1	(+5.4)	26.6	(-1.7)	33.6	(+2.4)
Quartz zinc, beige	40.9	(+6.2)	24.6	(-3.7)	32.8	(+1.6)
Quartz zinc, turquoise	42.6	(+7.9)	24.9	(-3.4)	34	(+2.8)
Quartz zinc, light blue	42.9	(+8.2)	24.7	(-3.6)	33.8	(+2.6)
Quartz zinc, natural, light grey	43.9	(+9.2)	26.4	(-1.9)	35.4	(+4.2)
Quartz zinc, natural, dark grey	45.7	(+11)	26.2	(-2.1)	35.8	(+4.6)
Ceramic tiles / panels 10 x 10 cm facing South					
White colour (glazed)	38.1	(+3.4)	22.7	(-5.6)	31.6	(+0.4)
Ochre colour	42.1	(+7.4)	23.4	(-4.9)	34.3	(+3.1)
Beige colour	42.6	(+7.9)	23	(-5.3)	34.1	(+2.9)
Light grey colour	43.3	(+8.6)	23.1	(-5.2)	34.4	(+3.2)
Brown colour	45.7	(+11)	24.4	(-3.9)	36	(+4.8)
Red-Dark red colour, black spots	45.8	(+11.1)	24	(-4.3)	36	(+4.8)
Olive green colour, black spots	46.6	(+11.9)	24	(-4.3)	36.1	(+4.9)
Dark grey colour	46.7	(+12)	23.5	(-4.8)	36	(+4.8)
Blue colour	47.1	(+12.4)	23.5	(-4.8)	36.3	(+5.1)
Black colour	47.1	(+12.4)	24.3	(-4)	36.5	(+5.3)
Black colour, with white spots	50.5	(+15.8)	24.3	(-4)	38	(+6.8)
Ceramic panels 25 x 13 x 1,3 cm facing South					
White colour	39.7	(+5)	23.2	(-5.1)	32.6	(+1.4)
Beige colour	40.8	(+6.1)	23.5	(-4.8)	33.3	(+2.1)
Light grey colour	41.8	(+7.1)	23.4	(-4.9)	33.6	(+2.4)
Brown colour	48.9	(+14.2)	22.6	(-5.7)	37	(+5.8)
Marble panels 20 x 30 cm facing South					
White colour	36.7	(+2)	22	(-6.3)	31.1	(-0.1)
Light grey colour	42	(+7.3)	22.3	(-6)	33.9	(+2.7)
Beige colour	44	(9.3)	21.3	(-7)	34.6	(+3.4)
Grey-Beige colour	45	(10.3)	22.3	(-6)	35.8	(+4.6)
Grey colour	46.3	(+11.6)	21.7	(-6.6)	36	(+3.4)
Ochre colour	50	(+15.3)	22.7	(-5.6)	37.8	(+6.6)
Dark grey colour	51.7	(+17)	22.7	(-5.6)	39.7	(+8.5)
Granite slabs 20 x 30 cm facing South					
White-Beige colour	39.3	(+4.6)	22.3	(-6)	32.8	(+1.6)
Salmon colour	42	(+7.3)	23	(-5.3)	33.8	(+2.6)
Red-Black colour	47.7	(+13)	23	(-5.3)	36.5	(+5.3)
Black colour	49.7	(+15)	23.3	(-5)	37.7	(+6.5)
Grey colour	50.3	(+15.6)	23.3	(-5)	38.3	(+7.1)

Table 5. Overview of the experimental measurements on samples of building materials facing South.

4. Conclusions

4.1 The effect of orientation

For paving materials, the horizontal placement, in combination with the high sun altitude angles during the summer period and the large amounts of solar radiation result in maximum absorption of solar radiation and the overheating of the materials, depending on their colour and thermo physical properties. For facade materials, the orientation determines the angle of incidence of the solar rays, the duration of insolation, as well as the hours of the day when it occurs.

In the case of northern facades, the solar radiation, which is received throughout the summer period, is limited. Consequently, the surface temperatures of north-facing materials, as the in-situ measurements show, are lower than or equal to the air temperatures throughout the summer days.

The insolation of southern facades is largely affected by the high sun altitude angles (74,4° for Athens, at noon on June the 21st). As a result, most south-facing materials do not record very high absolute maximum surface temperatures in summer.

Eastern and western facades receive large sums of solar radiation at angles which are nearly normal to their surfaces. Western facades may record higher surface temperatures than eastern ones due to the fact that during the afternoon, the large sum of solar radiation, which impinge on western facades, coincide with the highest daily air temperatures.

4.2 The effect of colour

The material property that defines its behaviour under the effect of solar radiation is reflectivity, which depends mainly on its colour (Givoni, 1998). Light-coloured materials record lower surface temperatures than dark-coloured ones. This fact has been largely documented by many researchers (Givoni, 1998; Akbari, et al., 1992; Doulos, et al., 2004) and was also evident from the results of this work shown above.

However, an important concern regarding light-coloured paving materials, is the effect of soiling and weathering on colour and, therefore, on their thermal behaviour. This fact is evident from the comparison of the maximum and the mean surface temperatures of the in-situ and the experimental measurements carried out with light-coloured materials in this study.

4.3 The effect of thermo physical properties

Even though orientation (see 4.1) and colour (see 4.2) affect the surface temperatures of the materials exposed to solar radiation, so also their thermo physical properties. This fact is evident from the maximum surface temperatures and the daily temperature variation of certain groups of materials, such as loose, earthen materials, light-weight concrete and wood.

The high surface temperatures (Tables 1 and 2) of loose earthen materials directly depend on their thermo physical properties. Dry earth has low thermal conductivity (0.25 to 0.30 W/mK) and density (300 to 1600 kg/m³) (Oke, 1995) values, depending on the composition of the soil (Oke, 1995). The values for wood are similar (about 0.09 W/mK for softwood and 0.19 W/mK for hardwood) (Oke, 1995). Consequently, the ability of these materials to diffuse heat through their mass (Oke, 1995; Givoni, 1998) and their ability to store heat (Oke, 1995; Szokolay, 1980), which are expressed by the theoretical properties of diffusivity and admittance, respectively, are relatively low. As a result, earthen materials and wood have a rather contradictory thermal behaviour. They heat up considerably until noon, but cool down rapidly (3 to 4 ºC every half hour) after 13:00 to 14:00, with their surface temperatures dropping below the air temperature by sunset (19:30 to 20:00).

4.4 The effect of shading

The measurements in the in-situ study clearly demonstrate that the effect of shading is of utmost importance in the case of dark-coloured materials (e.g. asphalt, dark-coloured stone, etc.). The differences in the surface temperatures of exposed and shaded dark-coloured materials are very large, ranging from 19 to 28 ºC for the absolute maximum temperatures, and from about 13 to about 21 ºC for the mean temperatures. In comparison to the mean ambient temperatures, the mean surface temperatures of the exposed materials are approximately 10 to 18 ºC higher, whereas shaded materials are, most of the time, cooler than the air temperature. Finally, around sunset, exposed materials are 5 to 9 ºC warmer than the air temperature, while shaded materials are considerably cooler. A detailed report on the effect of shading on the surface temperatures of materials is given by Bougiatioti (2005).

4.5 The effect of vegetation cover and water surfaces

Vegetation significantly affects the overall environmental conditions that prevail in the city centres (Givoni, 1998) in two distinct, and equally important ways. Firstly, by shading the various materials (see 4.4) and secondly, by maintaining low surface temperatures through the process of transpiration. However, it should be noted that the surface temperatures of vegetative cover depend on the characteristics of the species, and mainly their ability to withstand drought. In this study, the grass surfaces measured temperatures according to their thickness and irrigation patterns. The temperature of surfaces of thick and regularly irrigated grass was lower than the environmental temperatures throughout the day. This was also the case for various surfaces of evergreen shrubs. On the contrary, a grass surface, which was fairly worn and dry, developed a maximum surface temperature of 45 ºC, which was approximately 10 ºC higher than the air temperature.

Water has a very positive contribution to the microclimate, because of its high heat capacity (1160 Wh/m³K (Oke, 1995)), and its ability to evaporate, which causes considerable air and surface temperature reduction with a simultaneous increase of relative humidity levels (Oikonomou, 2004). In this study, the temperatures of water surfaces (mainly fountains in two squares in Athens) measured were very low (mean maximum T=25,5 ºC) compared to the corresponding air temperatures.

5. Selection of "skin" materials based on their thermal behaviour

5.1 Selection of paving materials

Apart from all the afore-mentioned parameters, the overheating of paving materials depends on the geometrical characteristics of the urban open spaces. A preliminary effort to define those "worst-case scenario" instances, namely the instances of urban open spaces where full -or almost full- insolation occurs during the summer months, is presented in a previous study by the author Bougiatioti (2006). The cases of urban canyons with North-to-South direction and with NW-to-SE direction, where the materials of the horizontal surfaces were bound to overheat during the summer, comprise of height-to-width (referred to from now on as H/W) ratios from 0 to 0.5 (or 1). For canyons with NE-to-SW direction, including narrower canyons, have H/W ratio of 1 (or 1.5). Finally, the horizontal surfaces of urban configurations, whose main axis runs East-to-West are prone to overheating in the summer, even in the case of a H/W ratio of 2.

In all the above cases, the negative contribution to the urban microclimate and to thermal comfort is important, and material selection should, thus, aim for:

- Materials which will measure low surface temperatures, compared to the air temperature, namely light-coloured materials
- Materials shaded by permanent or temporary shading structures and/or vegetation (trees or climbing plants) (Bougiatioti, 2005)
- Porous materials combined with an appropriate system of periodic water-sprinkling (Bougiatioti, 2005)
- Vegetative cover (e.g. grass, shrubs)
- Water surfaces
- Materials with low heat capacity and diffusivity, such as dry soil, light-weight concrete, ceramic blocks and wooden boards, which overheat during noon, but cool down quickly during the afternoon.

The application of one or more of the above solutions depends on the character and the use of the urban space, as well as on its design principles. Issues of visual comfort should also be taken into consideration, as the choice of certain materials (i.e. materials with light colours and/or glossy surfaces) can cause glare.

Finally, the materials for urban space should be chosen according to its peculiarities, its character and its patterns of use. During summer days, there should be shaded areas, where the solar radiation is blocked and the surface temperatures of the materials remain low (Khandaker Shabir, 2000). Similarly, during winter and the intermediate seasons (spring and autumn), there should be areas with no obstruction for solar radiation access.

5.2 Selection of facade materials

The overheating of facade materials facing East and West mainly occurs in buildings, which form part of wide urban canyons with H/W ratios smaller than 0.5. (Bougiatioti, 2006) In more narrow canyons, the overshadowing of the facades by the opposite buildings prevents the materials from heating to high surface temperatures.

In general, the materials, which form the outer layer of eastern and western facades, may heat up to very high surface temperatures. For the above-mentioned reasons, the choice of materials for eastern and western facades should aim for:

- Materials, which record low surface temperatures compared to the air temperature, namely light-coloured materials or materials painted with "cool paints" (Pomerantz, et al., 1998; Heat Island Group, 2005; Synnefa, et al., 2006; Synnefa, et al., 2007).
- Materials, irrespective of their surface temperatures, shaded by permanent or temporary shading structures and/or vegetation (trees or climbing plants).
- Vegetative cover with climbing plants).

- Vertical water surfaces. (Oikonomou, 2005)
- Materials with a low heat capacity and diffusivity, such as composite wood, aluminium and copper panels, which may overheat during the morning, but cool down quickly afterwards.
- Materials placed at a distance from the walls (ventilated facade system).

The effect of the overheating of east and west-facing materials on the internal climate of buildings can be minimised with the use of thermal insulation. (Bougiatioti, 2007).

5.3 Conclusion

It should be noted that the application of one or more of the afore-mentioned proposals to urban open spaces and buildings facades mainly depends on their use and design. For instance, for building facades, it is important to examine the aesthetic result of the application of one or more solutions to street frontages with a certain orientation at the scale of one or more building blocks. A good example is the redesign of the southern facades of city streets with shading devices and solar systems (e.g. PV panels for wall cladding or for shading). Such an intervention could help not only to improve the urban microclimate and the interior climate of buildings, but also to educate the public on issues of sustainability and energy conservation.

Finally, it should be noted that the choice of surfacing materials for both horizontal and vertical city surfaces affects visual comfort conditions in the urban open spaces, as well as the possibility of taking advantage of daylight for the illumination of the interior spaces of buildings (Givoni, 1993). Consequently, for narrow urban streets, where solar radiation penetrates for a limited period during the day, the use of light-coloured materials can help improve visual comfort conditions in the interior spaces of the buildings. (Yannas, 2001) On the contrary, in wide urban canyons with a small height-to-width (H/W) ratio, the use of light-coloured materials can cause glare and degrade visual comfort conditions at the level of pedestrian circulation.

6. Acknowledgements

The research, which is presented in this chapter, formed part of a Ph.D. thesis, which was supported by the Greek State Scholarships Foundation (I.K.Y.).

The parts of the paper referring to façade materials, as well as Tables 3, 4 and 5 were reprinted from J. Solar Energy, 83, Bougiatioti, F., Evangelinos, E., Poulakos, G. And Zacharopoulos, E., The summer thermal behaviour of "skin" materials for vertical surfaces in Athens, Greece as a decisive parameter for their selection, 582-598, Copyright (2008), with permission from Elsevier.

7. References

Akbari, H., Davis, S., Dorsano, S., Huang, J. & Winnett, St., (Eds.) (1992). *Cooling Our Communities*, U.S. E.P.A., ISBN 0-16-036034-X, Washington, U.S.A.

Atturo, C. and Fiumi, L. (2005). Thermographic analyses for monitoring urban areas in Rome to study heat islands, In: *Proceedings of 1st International Conference. Palenc 2005*, Santamouris M. (Ed.), vol. 2, Heliotopos Conferences Ltd, ISBN 960-88153-5-5, 960-88153-3-9, Athens, Greece, pp. 145-150.

Boon Lay, O., Guan Tiong, L. & Yu, C. (2000). A survey of the thermal effect of plants on the vertical sides of tall buildings in Singapore, In: *Proceedings of the 17th PLEA Conference*, Steemers K., Yannas S. (Eds.), James & James Science Publishers Ltd., ISBN 1-902916-16-6, London, UK, pp. 495-500.

Bougiatioti, F. (2005). The effect of shading on the surface temperatures of the materials used in the skin of greek cities. In: *Proceedings of 1st International Conference. Palenc 2005*, Santamouris, M. (Ed.), vol. 2, Heliotopos Conferences Ltd, ISBN 960-88153-5-5, 960-88153-3-9, Athens, Greece, pp. 775-780.

Bougiatioti, F. (2005). The effect of water-sprinkling on the surface temperatures of the materials used in the skin of greek cities, In: *Proceedings of 1st International Conference. Palenc 2005*, Santamouris, M. (Ed.), vol. 2, Heliotopos Conferences Ltd, ISBN 960-88153-5-5, 960-88153-3-9, Athens, Greece, pp. 749-754.

Bougiatioti, F. (2006). The effect of urban geometry on the surface temperatures of materials' used in the "skin" of greek cities, In: *Proceedings of PLEA 2006. 23rd International Conference*, Compagnon R., Haefeli P., Weber W., (Eds.), vol. 1, PLEA, HesSo, Universite de Geneve, ISBN 2-940-156301, Geneva, Switzerland, pp. 471-476.

Bougiatioti, F., Evangelinos, E., Poulakos, G. & Zacharopoulos, E. (2009). The summer thermal behaviour of "skin" materials for vertical surfaces in Athens, Greece as a decisive parameter for their selection. *J. Solar Energy*, 83, pp. 582-598.

Cadima, P. (1998). The effect of design parameters on environmental performance of the urban patio: a case study in Lisbon, In: *Proceedings of. PLEA 98*, Maldonado, Yannas S. (Eds.), James & James Science Publishers Ltd., ISBN 873936818, London, UK, pp. 171-174.

Cantuaria, G.A.C. (2000). A comparative study of the thermal performance of vegetation on building surfaces, In: *Proceedings of the 17th PLEA Conference*, Steemers K., Yannas S. (Eds.), James & James Science Publishers Ltd., ISBN 1-902916-16-6, London, UK, pp. 312-313.

Cook, J., Bryan, H., Agarwal, V., Deshmukh, A., Kapur, V. & Webster, A. (2003). Mitigating the Heat Impact of Outdoor Urban Spaces in a Hot Arid Climate, In: *Proceedings of 20th PLEA International Conference*, Bustamante W., Collados, E. (Eds.), ISBN (-), Santiago, Chile, E-16 [CD-ROM].

Doulos, L., Santamouris, M. & Livada, I. (2004). Passive Cooling of open urban areas. The role of materials, *J. Solar Energy*, 77, pp. 231-249.

Eymorfopoulou, A. & Aravantinos, D. (1998) The contribution of a planted roof to the thermal protection of buildings in Greece, *J. Energy and Buildings*, 27, 1, pp. 29-36.

Givoni, B. (1998). *Climate Considerations in Building and Urban Design*, Van Nostrand Reinhold, ISBN 0442009917, New York, U.S.A.

Heat Island Group, 2005, Available from:
<http://eetd.lbl.gov/Heatisland /Pavements/LowerTemps/>
Hoyano, A. (1988). Climatological uses of plants for solar control and the effects on the thermal environment of a building, *J. Energy and Buildings*, 11, 3, pp. 181-199.
Khandaker Shabir, A. (2000). Comfort in urban spaces: defining the boundaries of thermal comfort of outdoor thermal comfort for the tropical environments, In: *Proceedings of the 17th PLEA Conference*, Steemers K., Yannas S. (Eds.), Cambridge, 2000, James & James Science Publishers Ltd., ISBN 1-902916-16-6, London, UK, pp. 571-576.
Labaki, L.C., et al. (2003). The effect of pavement materials on thermal comfort in open spaces, In: *Proceedings of 20th PLEA International Conference*, Bustamante W., Collados, E. (Eds.), ISBN (-), Santiago, Chile, E-5. [CD-ROM]
Labaki, L.C., Oliveira M.C.A. & Freire A.P. (2003). The effect of pavement materials on thermal comfort in open spaces, In: *Proceedings of 20th PLEA International Conference*, Bustamante W., Collados, E. (Eds.), ISBN (-), Santiago, Chile, E-5. [CD-ROM]
Markus, T.A. & Morris, E.N. (1980). *Buildings, Climate and Energy*, Pitman Publishing Ltd., ISBN 0-273−00268-6, London, UK.
Marques de Almeida, D. (2002). Pedestrian Streets. Urban design as a tool for microclimate control, In: *Proceedings of PLEA 2002. Design with the Environment*, GRECO and ACAD (Eds.), vol. 1, ISBN (-), Toulouse, France, pp. 437-439.
Oikonomou, A. (2004). *Water in the city of Athens. An environmental and bioclimatic approach*, In: Ecological Design for an Effective Urban Regeneration, Babalis, D. (Ed.), Firenze University Press, ISBN 88-8453-179-9, Florence, Italy, pp. 129-136.
Oikonomou, A. (2005). Bioclimatic design of water elements in mediterranean cities, In: *ECOPOLIS: Sustainable Planning and Design Principles*, Babalis, D. (Ed.), Alinea Editrice, ISBN 88-6055-006-8, Florence, Italy, pp. 125-132.
Oke, T. R. (1995). *Boundary Layer Climates*, Routledge, ISBN 0-415-04319-0, London and New York.
Papadakis, G., Tsamis P. & Kyritsis S. (2001). An experimental investigation of the effect of shading with plants for solar control of buildings, *J. Energy and Buildings*, 33, pp. 831-836.
Pomerantz, M., et al. (1996). *Paving materials for heat island mitigation*, LBNL Report, LBL-38074, cited in Rosenfeld, A.H., Akbari, H., Romm, J.J. & Pomerantz, M. (1998). Cool communities: strategies for heat island mitigation and smog reduction, J. *Energy and Buildings*, 28, 1, pp. 51-62.
Santamouris, M. (Ed.) (2001). *Energy and Climate in the Urban Built Environment*, James and James Science Publishers Ltd., ISBN 1-873936-90-7, London, UK.
Synnefa, A., Santamouris, M. & Livada, I. (2006). A study of the thermal performance of reflective coatings for the urban environment, *J. Solar Energy*, 80, 8, pp. 968-981.
Szokolay, S. V. (1980). *Environmental Science Handbook*, The Construction Press, ISBN 0-86095-813-2, Lancaster / London / New York.

Yannas, S. (2001). Bioclimatic urban design principles, In: *Environmental Design of Cities and Open Spaces*, vol. 1, Amourgis, S. (Ed.), Hellenic Open University, ISBN 960-538-311-X, Patras, Greece, pp. 175-234.

8

Safe Drinking Water Generation by Solar-Driven Fenton-Like Processes

Benito Corona-Vasquez*, Veronica Aurioles and Erick R. Bandala
Grupo de Investigación en Energía y Ambiente, Fundación Universidad de las Américas,
Puebla, Santa Catarina Mártir, Cholula
México

1. Introduction

Around the world, it is estimated that 1.2 billion people have limited or no access to safe water for domestic use. As a result, the prevalence of water borne diseases affects not only the health of the inhabitants of these regions, but also their economic development (Gelover et al., 2006). It is clear that water resources management is becoming a critical issue worldwide, especially in regions with low rainfall and growing population (Bandala et al., 2010).

Just considering Africa, Latin America and the Caribbean, about one billion people have no access to safe water supplies resulting in serious human health effects, i.e. 1.5 million children died every year due to water borne diseases. Moreover, the lack of safe drinking water has been related to poverty and considerable limitations for sustainable development (Montgomery and Elimelech, 2007). In Mexico, for example, waterborne diseases affect over 6% of the total population, being rural communities with less than 2,500 inhabitants the most affected, since only 78% of this rural population has access to piped water (CONAGUA, 2011). Unfortunately, this situation is not limited to Mexico, but it is common in other developing countries in Latin America.

It is well known that human pathogens become sensitive to different environmental conditions once discharged into a water body. For example, some environmental conditions such as temperature and ultraviolet (UV) radiation are capable of inactivating waterborne pathogens. However, engineering processes are required if the goal is to assure the generation of safe drinking water for remote, poor, isolated regions in developing countries (Castillo-Ledezma et al., 2011).

Several different disinfecting processes have been tested in order to deactivate undesirable microorganisms in water. Within this variety of disinfectants, Advanced Oxidation Processes (AOPs) have been proven efficient and cost-effective for water treatment (Blanco et al., 2009). These physical-chemical processes have the potential of producing deep changes in the chemical structure of pollutants as a result of the action of hydroxyl radicals (HO•) (Orozco et al., 2008). Several scientific studies suggest that AOP's high efficiency is related to their thermodynamic viability and the increased rate of reaction produced by

---

* Corresponding Author

hydroxyl radicals. Among various AOPs, photocatalytic processes are very attractive for the mineralization (conversion to carbon dioxide, water, and other mineral species) of aqueous pollutants and inactivation of pathogenic microorganisms (Gelover et al., 1999; Bandala and Estrada, 2007; Bandala et al., 2007 & 2008). The use of AOPs for water disinfection, using solar radiation as the energy source, usually referred to as *enhanced photocatalytic solar disinfection* (ENPHOSODIS), has allowed the efficient deactivation of highly resistant microorganisms. Specifically, heterogeneous and homogeneous photocatalysis are the AOPs with the most technological applications because of their ability to remove organic pollutants and their capability to inactivate nuisance microorganisms. Regarding heterogeneous photocatalysis, the use of titanium dioxide (TiO2) as a catalyst has been widely tested and proven effective for deactivating several microorganisms as well as carcinogen cells (Dunlop et al., 2008; Rincon and Pulgarin, 2003; Reginfo et al., 2008; Castillo-Ledezma et al., 2011). In comparison with the heterogeneous arrangement, homogeneous processes also rely on the generation of hydroxyl radicals. Nevertheless, it has been proposed that other highly oxidant species could be involved in pollutant degradation and microorganisms deactivation. Fenton and Fenton-like processes are among the most widely studied methodologies (Bandala et al., 2009; Guisar et al., 2007; Bandala et al., 2011). From the economic point of view, the possibility of using solar energy to promote both homogeneous and heterogeneous photocatalytic processes is an interesting alternative to the use of these technologies in developing countries (Bandala et al., 2011; Blanco et al., 2007).

The aim of this chapter is to review the state-of-the-art in the use of solar driven Fenton-like processes for the deactivating waterborne pathogens. It also the goal of this work to discuss the advantages and potential limitations of these treatment processes while analyzing the challenges and opportunities for the application of such technologies at real scale in poor, isolated regions in developing countries with no access to safe drinking water.

2. Chemical and biological mechanisms involved in homogeneous photocatalysis

2.1 Chemistry

The chemical mechanisms involved in the Fenton reaction are well known since early of the past century. The reactions of iron (II) salts with hydrogen peroxide have been widely studied for decades and the main reactions involved are summarized in Table 1 (Orozco et al., 2008; Gallard and De Laat, 2000; Gallard et al., 1999).

$Fe^{2+} + H_2O_2 \rightarrow Fe^{3+} + HO^{\bullet} + OH^{-}$	(1)
$Fe^{2+} + HO^{\bullet} \rightarrow Fe^{3+} + OH^{-}$	(2)
$Fe^{3+} + H_2O \leftrightarrow [FeOH]^{2+} + H^{+}$	(3)
$Fe^{3+} + H_2O_2 \leftrightarrow [FeHO_2]^{2+} + H^{+}$	(4)
$[FeOH]^{2+} + H_2O_2 \leftrightarrow [Fe(OH)(HO_2)]^{+} + H^{+}$	(5)
$[FeOH]^{2+} \rightarrow Fe^{2+} + HO^{\bullet}$	(6)
$[FeHO_2]^{2+} \rightarrow Fe^{2+} + HO_2^{\bullet}$	(7)
$[Fe(OH)(HO_2)]^{+} \rightarrow Fe^{2+} + HO_2^{\bullet} + HO^{-}$	(8)
$RH + HO^{\bullet} \rightarrow R^{\bullet} + H_2O$	(9)

Table 1. Chemical reactions involved in the Fenton reaction (Fe (II) and H_2O_2).

When the process is carried out under radiation, called the photo-Fenton process, the influence of radiation increases the oxidation rate of the pollutant when compared to the conventional Fenton reaction (Bandala et al., 2007). Several possible reaction schemes for the photo-induced system have been proposed as well as the formation of different complexes studied by spectrophotometry. From these studies, the influence of iron oxidation state and the specific counter ion of the iron salt employed have been demonstrated to affect both the decomposition of H_2O_2 as well as the overall efficiency of the photo-Fenton process for the oxidation of some model pollutants.

$$Fe^{2+} + Cl^- \leftrightarrow FeCl^+ \tag{10}$$
$$Fe^{3+} + Cl^- \leftrightarrow FeCl_2^+ \tag{11}$$
$$Fe^{3+} + 2Cl^- \leftrightarrow FeCl_2^+ \tag{12}$$
$$FeCl_2^+ \rightarrow Fe^{2+} + Cl^\bullet \tag{13}$$
$$FeCl^{2+} \rightarrow Fe^{2+} + Cl^{\bullet\,2-} \tag{14}$$
$$Fe^{2+} + Cl^\bullet \rightarrow Fe^{3+} + Cl^- \tag{15}$$
$$Fe^{2+} + Cl_2^\bullet \rightarrow FeCl_2^+ + Cl^- \tag{16}$$
$$Cl^- + HO^\bullet \rightarrow [ClOH]^{\bullet-} \tag{17}$$
$$[ClOH]^{\bullet-} + H^+ \rightarrow [HClOH]^\bullet \tag{18}$$
$$[HClOH]^\bullet \rightarrow Cl + H_2O \tag{19}$$
$$Cl^\bullet + H_2O_2 \rightarrow HO_2^\bullet + Cl^- + H^+ \tag{20}$$
$$Cl_2^\bullet + H_2O_2 \rightarrow HO_2^\bullet + 2Cl^- + H^+ \tag{21}$$

Table 2. Suggested side reactions related with the iron salt counter ion during Fenton reaction.

2.2 Biological mechanisms

In general, photocatalytic processes, both homogeneous and heterogeneous, in the presence of iron and hydrogen peroxide have been demonstrated effective against a wide variety of resistant microorganism such as viruses (Kim et al., 2010), helminth eggs (Bandala et al., 2011a,b; Guisar et al., 2007), bacteria and spores (Dunlop *et al.*, 2008; Bandala *et al.*, 2009; Bandala et al., 2011c; Sichel *et al.*, 2009; Castillo-Ledezma et al., 2011).

The main mechanism involved in deactivating pathogenic microorganisms is suggested to be related with the cellular damage produced by so-called reactive oxygen species (ROS), mainly hydroxyl (HO$^\bullet$) and superoxide radical (O$_2^\bullet$) as shown in Tables 1 and 2. According to different studies, these ROS are able to modify and eventually destroy the structure of the cell membrane (Alrousan *et al.*, 2009, Malato *et al.*, 2009), mainly as the result of lipid peroxidation (Dunlop *et al.*, 2008; Alrousan *et al.*, 2009). The initial damage is produced in the outer lipopolisaccarid and peptidoglycan walls, followed by lipid peroxidation and protein and polysaccarides oxidation (Malato *et al.*, 2009; Dalrymple *et al.*, 2010) affecting the regulatory function of the cell membrane for the internal and external interchange. The damage produced will further produce failure in the cell's respiratory activity and decrease its permeability, allowing the attack of inner cell components leading to its death (Alrousan *et al.*, 2009).

Some studies have also demonstrated that microorganism's deactivation is also improved by the presence of iron derivatives, which have been suggested to show an important inhibitory activity in important microbiological processes such as biofilm generation

(Dunlop *et al.*, 2008). Cells are used to regulate iron adsorption as a defense mechanism against hydroxyl radical; however once hydroxyl radicals are generated in the intracellular media, as a result of the Fenton-like process by direct attack of the ROS, they are free for reacting with biomolecules (Darlymple *et al.*, 2009).

During cellular metabolism some ROS are produced, such as superoxide ($O_2^{\bullet}$), hydroxyl radical and hydrogen peroxide (H_2O_2), as a result of cell respiration. However, these oxidizing species are in equilibrium with the immune system defense mechanism through anti-oxidizing enzyme production related to superoxide dismutase (SODs), catalase (CAT) and glutathione peroxidase (GPX) families (Castillo-Ledezma et al., 2011). When microorganisms are exposed to a major oxidative stress, for example ROS produced during a photocatalytic process, enzyme production is no longer capable of eliminating excess radical allowing deep cell damage. In the same way, ROS may produce additional oxidative stress in the cells through Fenton and Heber-Weiss reaction (Dunlop *et al.*, 2008) generating damage in all the cell components including proteins, lipids and DNA. In the case of DNA damage, produced by pyrimidine dimmers formation by the generation of covalent bonds among the bases in the same DNA chain (Sichel *et al.*, 2009), it generates mutations that may lead to loss of functional capability and death of cell (Malato *et al.*, 2009). At the same time, when microorganisms are exposed to ultraviolet radiation (UV, $\lambda \leq 400$ nm) during the photocatalytic reaction, DNA damage occurred directly through the radiation absorption by cell chromophores, which absorb radiation and produce heat. This interaction leads to an increase in ATP and RNA synthesis, jointly with the increase of ROS production. Microorganisms receiving a sub-lethal dose of UV radiation may become resistant to induced oxidative stress, partially recover their defense mechanisms and adapt to oxidative stress generated by exposure to UV radiation alone (tanning effect) (Bandala et al., 2011b).

3. Solar-driven Fenton and Fenton-like inactivation of pathogenic microorganisms

Homogeneous processes rely on the generation of hydroxyl radicals. Nevertheless, it has been proposed that other highly oxidant species could also be involved in pollutant degradation (Anipsitakis and Dionysiou, 2004). Fenton and Fenton-like processes are among the most widely studied methodologies. When a Fenton process uses ultraviolet (UV) radiation, visible light or a combination of both, the resulting process (known as photo-Fenton) has several advantages, including the increase of degradation rate and the flexibility of using alternative energy sources (i.e., solar radiation) for driving the process (Bandala et al., 2007; Bandala and Estrada, 2007; Fernandez et al., 2005). From the economic point of view, the possibility of using solar energy to promote both homogeneous and heterogeneous photocatalytic processes is an interesting alternative technology for use in developing countries (Bolton, 2001).

Several reports on the use of photo-Fenton process for deactivating pathogenic microorganisms have been published in the near past. In these studies, a wide variety of analysis have been conducted and reported, such as: the effects of the iron salt and pH on deactivating *E. coli* (Spuhler et al., 2010), the capability of solar driven photo-Fenton process to achieve simultaneous degradation of natural organic matter (NOM) and water disinfection (Moncayo-Lasso et al., 2009) as well as the effect of many other specific parameters as reported in a recent review by Malato et al. (2009).

In addition to the study of *E. coli* cells, many other microorganisms have been tested and used as indicators for evaluating the performance of photo-Fenton processes such as *Salmonella spp.* (Sciacca et al., 2011), *Fusarium solani* spores (Polo-Lopez et al., 2011), helminth eggs (Bandala et al., 2011a,b) and *Bacillus subtilis* spores (Bandala et al., 2011c). In a recent work, several different Fenton reagent concentrations were tested in combination with UV-A radiation (λ_{max}= 365 nm) to pursue deactivation of *B. subtilis* spores. The best spore deactivation conditions were found using [Fe(II)] = 2.5 mM and [H_2O_2] = 100 mM and UV-A radiation. As depicted in Figure 1, under these experimental conditions, over a 9-log reduction in spore viability was reached after 20 minutes of reaction. Interesting results were also observed from experiments conducted with low Fe(II) concentrations or even when no Fe(II) was added and only H_2O_2 and UV-A radiation were used. Under these experimental conditions, a lag phase –where no deactivation occurred- was observed during earlier stages of the disinfection process and much lower spore viability was determined after long time of irradiation. These results might suggest that microorganisms have the capability to generate defense mechanisms as a response to threatening environmental stresses. It is also suggested that the observed initial delays in the inactivation process may be due to the effect of defense mechanisms by the microorganisms against low ROS concentrations generated under these reaction conditions.

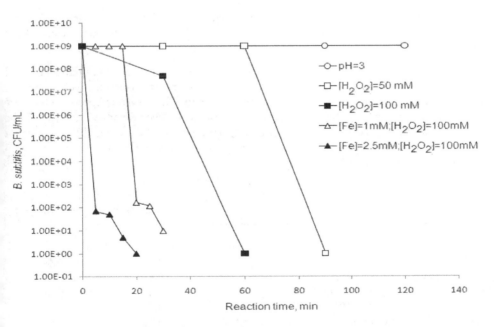

Fig. 1. Deactivation of *Bacillus subtilis* spores using photo-assisted Fenton reaction.

Furthermore, researchers have hypothesized that, when the iron salt is added to the reaction mixture, ROS generation increases dramatically and overwhelms the defense capability of the microorganisms leading to their immediate death without undergoing the lag phase observed previously.

The effects of ionic strength and natural organic matter (NOM) in spore deactivating kinetics have also tested and representative experimental results are illustrated in Figures 2 and 3, respectively. In both cases, an important decrease on the deactivating rate was observed when NOM concentration and ionic strength were increased. Experimental data has been modeled using a modification of the delayed Chick-Watson model, including the accumulated energy (Q_n), rather than the traditional C×t factor (disinfectant concentration times contact time) used for chemical disinfection (Bandala et al., 2009). Model fitting parameters, including deactivating rate constants for the different experimental conditions, are reported in Table 3.

$$\frac{N}{N_0} = \begin{cases} \dfrac{N}{N_0} \ if \ Q_n \leq Q_{nlag} = \dfrac{1}{k} \ln\left\{ \left(\dfrac{N_1}{N_0}\right)\left(\dfrac{N_0}{N}\right)_c \right\} \\[2ex] \dfrac{N_1}{N_0} e^{-k_1 Q_n} \ if \ Q_{nlag} \leq Q_n \leq Q_{n2} \\[2ex] \dfrac{N_2}{N_0} e^{-k_2 Q_n} \ if \ Q_n \geq Q_{n2} = \dfrac{1}{k_2 - k_1} \ln\left(\dfrac{N_2}{N_1}\right) \end{cases} \tag{22}$$

It was demonstrated that the lag-phase described initially for spore deactivation was avoided by the use of photo-assisted Fenton reaction whereas very different results were obtained when natural organic matter was present in raw water.

As observed in the experimental results, the photo-assisted Fenton reaction might represent an interesting alternative to deactivate recalcitrant microorganisms in water. In this particular study, the photo-assisted process was used to kill B. subtilis spores, which are currently considered among the most resistant bacteria. It has been demonstrated that, if the photo-Fenton reaction is capable to deactivate B. subtilis spores, it could be able to eliminate other less resistant pathogenic microorganisms present in water under the same reaction conditions (Bandala et al., 2011a).

Experimental conditions	k_2(min^{-1})	$\ln\left(\dfrac{N_1}{N_0}\right)$	$\ln\left(\dfrac{N_2}{N_0}\right)$
UV, pH = 3	-	-	-
[H$_2$O$_2$] = 50 mM	-	2.45	-
[H$_2$O$_2$] = 100 mM	0.39	0	2.99
[Fe(II)] = 1mM; [H$_2$O$_2$] = 100 mM	0.83	2.53	15.23
[Fe(II)] = 2.5 mM; [H$_2$O$_2$] = 100mM	0.87	2.78	15.67
[Fe(II)] = 2.5 mM; [H$_2$O$_2$] = 100 mM; [Cl$^-$] = 25 mgL^{-1}	0.85	2.72	15.82
[Fe(II)]=2.5 mM; [H$_2$O$_2$] = 100 mM; [Cl$^-$] = 50 mgL^{-1}	0.02	2.77	16.61
[Fe(II)] = 2.5 mM; [H$_2$O$_2$] = 100 mM; [Cl$^-$] = 100 mgL^{-1}	0.02	2.69	16.11
[Fe(II)] = 2.5 mM; [H$_2$O$_2$] = 100 mM; [SR-NOM] = 2.5 mgL^{-1}	0.74	2.53	17.05
[Fe(II)] = 2.5mM; [H$_2$O$_2$] = 100 mM; [SR-NOM] = 5.0 mgL^{-1}	0.76	2.3	16.12

Table 3. Kinetic data obtained using Chick-Watson model for homogeneous photocatalytic disinfection.

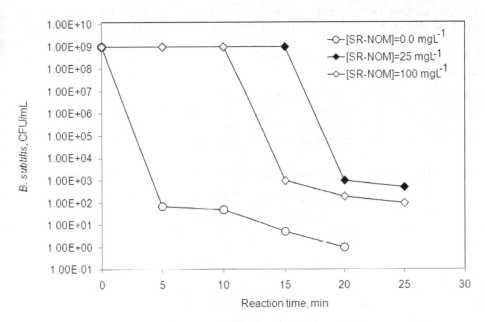

Fig. 2. Effect of natural organic matter (NOM) on the efficiency of photo-assisted deactivation of *B. subtilis* spores.

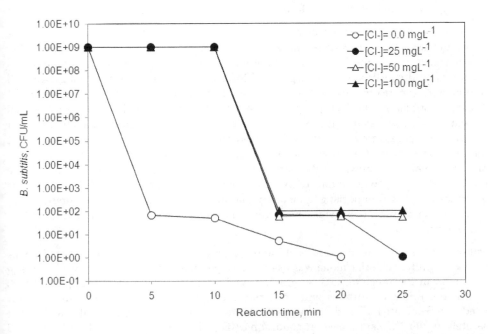

Fig. 3. Effect of ionic strength on the efficiency of photo-assisted deactivation of *B. subtilis* spores.

Spore deactivation using photo-Fenton reaction is considerably affected by the ionic strength and natural organic matter, mainly by delaying the beginning of actual deactivation process. As shown in Figure 2, an increase in the concentration of natural organic matter has an important slowing effect on spore deactivation since it increases the duration of the lag phase.

The experimental results presented in Figures 2 and 3 suggest that the efficiency of photo-Fenton processes used for disinfection depend largely on the quality of background water. Therefore, if these processes are to be cost-effective, they should be coupled with pre-treatment and/or other conventional drinking water processes.

4. Sequential disinfection using Solar-driven Fenton-like processes

As proposed earlier, AOPs by themselves may not be capable enough to reach the desired level of pathogen deactivation as well as to assure quality of treated water because of their lack of disinfectant residual. In the past, different authors have suggested that the application of a strong oxidant (i.e. ozone) followed by a weaker oxidant (i.e. free or combined chlorine) could produce important synergistic effects on the deactivating kinetics of strong pathogens such as *Cryptosporidium parvum* oocysts (Rennecker et al., 2000; Driedger et al., 2000; 2001). The high efficiency demonstrated by sequential disinfection using the ozone-chlorine pair could be related to the generation of different reactive oxygen species (i.e. hydroxyl radicals) by the use of ozone which could synergically enhance the oxidative properties of chlorine, improving the overall inactivation rate. Considering this hypothesis, it could be possible that other methods producing hydroxyl radicals would be able to produce similar synergistic effects in similar sequential processes.

In a recent work, the efficiency of the sequential application of AOPs followed by free chlorine processes was investigated in order to deactivate helminth eggs, another highly resistant waterborne pathogen which is commonly found in surface waters contaminated with untreated domestic wastewater. Detailed information on the experimental matrix and methods used to perform these tests and to evaluate microorganism viability are described elsewhere (Bandala et al., 2011b). Figure 4 depicts the experimental results of sequential deactivation of *Ascaris suum* eggs using solar photocatalysis followed by free chlorine. For comparison purposes, notice that the inactivation results of *A. suum* eggs using free chlorine alone are also shown. It is important to mention that the reaction time, rather than the common CT value, is shown on the horizontal axis since the initial concentration of free chlorine (7.0 mg/L) was found to remain constant throughout the duration of the experiments. It is also important to mention that all deactivating tests were conducted in synthetic water at pH 7.0, at which hypochlorous acid (HOCl) was the predominant free chlorine species.

As shown in the experimental results of Figure 4, practically no effect on egg viability was observed when free chlorine alone was used as a single disinfectant. On the other hand, egg deactivation in the range of 25-30% was achieved when free chlorine was applied for 20 minutes after primary treatment with solar photocatalysis was applied. Experimental runs labeled as "Sequential 1" and "Sequential 2" are two different experiments carried out separately under the same reaction conditions ([Fe(II)]= 5mM; [H_2O_2]= 140 mM after 60 minutes of photo-assisted treatment). As mentioned, the effect of chlorine by itself on egg viability is almost negligible. In a previous work, Bandala et al. (2011b), reported that comparing the effect of chlorine alone with the photo-assisted process alone, the sequential process shows a very close trend to the observed for the AOP. Nevertheless, after the initial 20 minutes of application of sequential chlorine deactivation, the eggs viability was observed remaining unchanged at 10%, whereas in the case of the photo-Fenton process, the helminth eggs deactivation continues until reaching almost 2-log deactivation.

The use of highly-resistant pathogens, such as helminth eggs, as a conservative surrogate for water disinfection is also a very interesting issue because the deactivation of the helminth eggs is a complex task, which was in agreement with previous results from our research group (Bandala et al., 2011b), achieving almost complete helminth eggs deactivation.

Any other less resistant pathogen microorganisms (i.e. bacteria) present in the raw water will be deactivated under the same reaction conditions and after providing the same solar radiation dose. It is reasonable then that the disinfection level reached using AOPs may be, as demonstrated here, improved by the further adding of free chlorine in the sequential process.

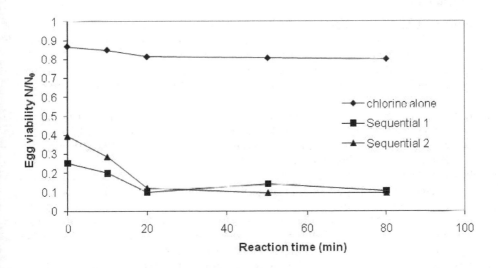

Fig. 4. Effect of sequential AOPs-chlorine processes on the deactivation of *Ascaris suum* eggs.

5. Conclusions and final comments

The use of solar radiation to catalyze Fenton-like processes has been proven to be effective to deactivate waterborne pathogens. Experimental results and reported deactivation parameters suggest that these processes could be an interesting alternative to conventional chlorine disinfection for developing countries. Furthermore, solar-driven Fenton-like processes is a more efficient alternative when compared to conventional solar disinfection (SODIS), since the required exposure times for both are significantly different and much lower for the photocatalytic arrangement. Nevertheless, it is well documented that for solar photocatalysis to perform at optimum conditions, a primary treatment of raw water might be needed in order to decrease the concentrations of interfering compounds such as natural organic matter. When considering the necessity to provide a disinfectant residual in order to protect water treated with AOPs, sequential disinfection using AOPs as a primary process coupled with secondary free chlorine also represents an interesting alternative for developing countries. Similar to the synergistic effects produced by ozone on secondary free chlorine deactivation of strong microorganisms such as *C. parvum* oocysts, preliminary studies conducted with AOPs used as primary disinfectant –instead of ozone- have reported to generate similar synergistic effects on the deactivation of conservative surrogates such as *B. subtilis* spores and helminth eggs. More comprehensive work needs to be conducted in order to fully characterize the magnitude and occurrence of this synergism under a wide range of experimental conditions of interest for drinking water treatment such as water pH and temperature. Finally, in order to fully understand the effectiveness of solar-driven Fenton-like processes when used for disinfection purposes, it is clear that more studies are required in order to elucidate the actual mechanism of cell deactivation when using such processes.

6. References

Alrousan, D.M.A., Dunlop, P.S.M., McMurray, T.A., Byrne, A. Photocatalytic inactivation of *E. coli* in surface water using immobilised nanoparticle TiO_2 films. Water research 43 (2009) 47-54.

Anipsitakis G.P., Dionysiou D.D. 2004. Transition metal/UV-based advanced oxidation technologies for water decontamination. Applied Catalysis B: Environmental, 54, 155-163.

Bandala E.R., Pelaez M.A., Dionysiou D.D., Gelover S., García A.J., Macías D. 2007. Degradation of 2,4-dichlorophenoxyacetic acid (2,4-D) using cobaltperoximonosulfate in Fenton-like process. Journal of Photochemistry and Photobiology A: Chemistry 186, 357–363.

Bandala E.R., Estrada C. 2007. Comparison of solar collection geometries for application to photocatalytic degradation of organic contaminants. Journal of Solar Energy Engineering 129, 22–26.

Bandala E.R., Pelaez M.A., Garcia-Lopez J., Salgado M.J., Moeller G. 2008. Photocatalytic decolourization of synthetic and real textile wastewater containing benzidine-based azo dyes. Chemical Engineering and Processing 47, 169–176.

Bandala E.R., Corona-Vasquez B., Guisar R., Uscanga M. 2009. Deactivation of highly resistant microorganisms in water using solar driven photocatalytic processes. International Journal of Chemical Reactor Engineering 7, A7.

Bandala E.R., González L., de la Hoz F., Pelaez M.A., Dionysiou D.D., Dunlop P.S.M., Byrne J.A., Sanchez J.L. 2011a. Application of azo dyes as dosimetric indicators for enhanced photocatalytic solar disinfection (ENPHOSODIS). Journal of Photochemistry and Photobiology A: Chemistry 218, 185-191.

Bandala E.R., Gonzalez L., Sanchez-Salas J.L., Castillo J.H. 2011b. Inactivation of Ascaris eggs in water using sequential solar driven photo-Fenton and free chlorine. Journal of Water and Health, (In press).

Bandala E.R., Duran J.A., Holland J.N. 2011c. Consequences of Global Climate Change for Water Quality and Community Sustainability along the U.S.-Mexico Trans-Border Region: Case Studies of Reynosa/McAllen and Laredo/Nuevo Laredo. Proceedings of the Puentes Consortium's Mexico-U.S. Higher Education Leadership Forum. March 2011 Rice University, Houston, Texas. Available online at http://www.puentesconsortium.org/papers/working-papers.

Blanco J., Malato S., Fernandez-Ibañez P., Alarcon D., Gernjak W., Maldonado M.I. 2009. Review of feasible solar energy applications to water processes, Renew. Sustain. Energ. Rev. 13 (6-7), 1437-1445.

Blanco J., Fernandez P., Malato S. 2007. Solar photocatalytic detoxification and disinfection of water: an overview. Journal of Solar Energy Engineering 129 (1), 4-15.

Bolton J.R. 2001. Ultraviolet applications handbook. Bolton Photosciences Inc. Ontario, Canada.

Castillo-Ledezma J.H., Sanchez-Salas J.L., Lopez-Malo A., Bandala E.R. 2011. Effect of pH, solar irradiation, and semiconductor concentration on the photocatalytic disinfection of Escherichia coli in water using nitrogen-doped TiO_2. European Food Research Technology 233, 825-834.

Chacon J.M., Leal M.T., Sanchez M., Bandala E.R. 2006. Solar photocatalytic degradation of azo-dyes by photo-Fenton process. Dyes Pigments 69, 144-150.

Comision Nacional del Agua (CONAGUA). 2011. Estadísticas del agua en Mexico. Secretaría de Medio Ambiente y Recursos Naturales (SEMARNAT), Mexico. March, 2011.

Dalrymple, O.K., E. Stefanakos., M.A. Trotz., D.Y. Goswami. 2010. A review of the mechanisms and modeling of photocatalytic disinfection. Applied Catalysis B: Environmental 27-38.

Driedger, A.M., Rennecker, J.L. & Mariñas, B.J. 2001. Inactivation of Cryptosporidium parvum oocysts with ozone and monochloramine at low temperature. Wat. Res. 35(1), 41-48.

Driedger, A.M., Rennecker, J.L. & Mariñas, B.J. 2000. Sequential inactivation of Cryptosporidium parvum oocysts with ozone and free chlorine. Wat. Res. 34 (14), 3591-3597.

Dunlop P.S.M., McMurray T.A., Hamilton J.W.J., Byrne J.A. 2008. Photocatalytic-inactivation of Clostridium perfringens spores on TiO_2 electrodes. Journal of Photochemistry and Photobiology A: Chemistry 196, 113-119.

Fernandez P., Blanco J., Sichel C., Malato S. 2005. Water disinfection by solar photocatalysis using compound parabolic collectors. Catalysis Today 101, 345-352.

Gelover S., Gomez L.A., Reyes K., Leal T. 2006. A practical demonstration of water disinfection using TiO_2 films and sunlight. Water Research, 40 (17), 3274-3280.

Gelover S., Leal T., Bandala E.R., Román A., Jiménez A., Estrada C. 1999. Catalytic photodegradation of alkyl surfactants, Water Science and Technology 42 (5-6), 110-116.

Gallard H., De Laat J. 2000. Kinetic modeling of $Fe(III)/H_2O_2$ oxidation reactions in dilute aqueous solutions using atrazine as a model organic compounds. Water Research 34, 3107 - 3116.

Gallard H., De Laat J., Legube B. 1999. Spectrophotometric study of the formation of iron (III) – hydroperoxy complexes in homogeneous aqueous solutions. Water Research 33, 2929 – 2936.

Guisar R., Herrera M.I., Bandala E.R., García J.L., Corona B. 2007. Inactivation of waterborne pathogens using solar photocatalysis. Journal of Advanced Oxidation Technologies 10 (2), 1–4.

Haber F., Weiss J. 1934. The catalytic decomposition of hydrogen peroxide by iron salts. Proceedings of the Royal Society A 134, 332 – 351.

Kim, J.Y., Lee, Ch., Sedlak, D.L., Yoon, J., Nelson, K.L. 2010. Inactivation of MS2 coliphage by Fentons' reagent. Water Reserch 44, 2647-2653.

Malato S., Fernández-Ibáñez P., Maldonado M.I., Blanco J., Gernjak W. 2009. Decontamination and disinfection of water by solar photocatalysis: Recent overview and trends. Catalysis Today 147(1), 1-59.

Montgomery M.A., Elimelech M. 2007. Water and sanitation in developing countries: including health in the equation. Environmental Science and Technology 41 (1),17–24.

Moncayo-Lasso A., Sanabria J., Pulgarin C., Benítez N. 2009. Simultaneous E. coli inactivation and NOM degradation in river water via photo-Fenton process at natural pH in solar CPC reactor. A new way for enhancing solar disinfection of natural water. Chemosphere 77(2), 296-300.

Orozco S.L., Bandala E.R., Arancibia C.A., Serrano B., Suarez R., Hernández I. 2008. Effect of iron salt on the color removal of water containing the azo-dye reactive blue 69 using photo-assisted $Fe(II)/H_2O_2$ and $Fe(III)/H_2O_2$ systems, Journal of Photochemistry and Photobiology A: Chemistry 198, 144–149.

Polo-López M.I., García-Fernández I., Oller I., Fernández-Ibáñez P. 2011. Solar disinfection of fungal spores in water aided by low concentrations of hydrogen peroxide. Photochemistry and Photobiology Sciences 10, 381-388.

Reginfo-Herrera J.A., Mielczarski E., Mielczarski J., Castillo N.C., Kiwi J., Pulgarin C. 2008. Escherichia coli inactivation by N, S co-doped commercial TiO_2 powders under UV and visible light. Applied Catalysis B: Environmental 84 (3–4), 448–456.

Rennecker, J.L., Driedger, A.M., Rubin, S.A. Mariñas, B.J. 2000. Synergy in sequential inactivation of Cryptosporidium parvum with ozone/free chlorine and ozone/monochloramine. Water Research 34(17), 4121-4130.

Rincón A.G., Pulgarín C. 2003. Photocatalytical inactivation of E. coli effect of (continuous-intermitent) light intensity and of (suspended-fixed) TiO_2 concentration. Applied Catalysis 44, 263–284.

Sciacca F., Rengifo-Herrera J. A., Wéthé J., Pulgarin C. 2011. Solar disinfection of wild Salmonella sp. in natural water with a 18 L CPC photoreactor: Detrimental effect of non-sterile storage of treated water. Solar Energy 85(7), 1399-1408.

Sichel, C., Fernández-Ibáñez, P., de Cara, M., Tello, J. 2009. Lethal synergy of solar UV-radiation and H_2O_2 on wild Fusarium solani spores in distilled and natural water. Water Research 34, 1841-1850.

Spuhler D., Rengifo-Herrera J.A., Pulgarin C. 2010. The effect of Fe^{2+}, Fe^{3+}, H_2O_2 and the photo-Fenton reagent at near neutral pH on the solar disinfection (SODIS) at low temperatures of water containing Escherichia coli K12. Applied Catalysis B: Environmental 96(1-2), 126-141.

9

An Opaque Solar Lumber Drying House Covered by a Composite Surface

Kanayama Kimio[1], Koga Shinya[2], Baba Hiromu[3] and Sugawara Tomoyoshi[4]
[1]Kitami Institute of Technology
[2]Kyushu University
[3]Formerly, Kitami Institute of Technology
[4]Marusho-giken Co., Ltd.
Japan

1. Introduction

Since old time, natural air drying technique had been utilized in several areas of the world before artificial drying methods for wood were developed. In Hokkaido, Japan, a research on the modified type, in place of the natural air dryer, was developed as a practical technology. For instance, the solar dryers of greenhouse-type and solar house-type were made by hand or constructed architecturally. After tested at the public institute, some of the apparatus were distributed to some areas of northern part of Hokkaido. For several decades, from just after the first energy crisis 1973 until the recent time, as an application of solar radiation in the field of forest industry, an active-type and a semi passive-type of solar wood dryers have been developed experimentally and constructed annually. For instance, a semi passive-type handmade solar lumber dryer was invented by the Hokkaido Forest Industry Product Institute, which occupied one era in a wide area as an auxiliary lumber dryer. Recently, in order to suppress the global warming, a policy to reduce emission of carbon dioxide was urgently required on the worldwide. For example, an artificial steam-type lumber drying emits a lot of carbon dioxide because it consumes a lot of fossil oil, i.e., about fifty liters per one cubic meters of wood.

In this chapter therefore, in drying process of wood materials in forest industry, by improving the greenhouse-type dryer or solar house-type dryer, several highly advanced models of solar lumber drying apparatus were provided and structured. A test of performance of one of them was carried out successfully at the experimental site in Hokkaido. In this case, practical use of solar radiation under a new concept of transparent insulation-blackbody cavity effect, two models of a fully passive-type solar lumber drying house were designed and constructed. One of which is a south-north model and the other is an east-west model. The measurement and analysis on the working performance of both models were carried out by our project team for several years under severe winter season. The results obtained from both models were fairly good and compared well with each other. Consequently, from the performance, both models of the fully passive solar lumber drying house developed, in this national project, were recognized to be good for drying laminar lumber, even under a sever cold season in Hokkaido.

This research is based on a new concept, therefore, a special feature involved in the article, will be expressed by a few technical keywords which will introduce the contents as follows: (1) transparent insulation/blackbody cavity effect, (2) composite surface, (3) CF-sheet, (4) coefficient of transmittance-absorptance, (5) volumetric S. R. incidence, (6) efficiency of volumetric solar heat collection, (7) volumetric solar heat collected, (8) insulated cylinder (chimney), (9) thermo siphon effect, and so on.

1.1 Short historical overview on the related researches

For several decades until this time in Japan, the researches on a solar lumber drying apparatus are mainly as follows: (1) Firstly, a small sized semi passive-type solar lumber dryer made by hand was experimented by Hokkaido Forest Research Institute (Norota, T., et al., 1983) and some products were practically used over northern area in Hokkaido. (2) Second, a large scale solar lumber drying house, an active-type, was developed by a big company, as a national project (Miyoshi, M., Sep., 1987), (Kanayama, K., Baba, H., 2004), and was examined for three years. (3) Third, as a result of the above task, a larger one, using the same type of apparatus was constructed by technical transfer to overseas. In Indonesia, a much bigger size active-type solar lumber drying system was constructed for drying a broadleaf wood, like lauan, during several year (Yamada, M., 1998).

However, in this research aimed at solar lumber drying by improving an agricultural vinyl house, an active-passive type was examined (Kanayama, K., et al., 2006), adopting the new concept as above, and "a fully passive-type solar lumber drying" was ultimately created by our project team (Kanayama, K., et al., 2007), (Kanayama, K., et al., 2008), (Baba, H., et al., 2008). This technical article (Kanayama, K., et al., 2009), (Baba, H., et al. 2009), firstly deals with an optical-thermal mechanism of volumetric solar radiation (S. R.) incidence and a capability of volumetric solar heat collected into "a fully passive-type solar lumber drying house" covered by a composite film, consisting of a triple transparent film and a CF-sheet. The outside view of "a fully passive-type solar lumber drying house" looks like non-transparent from the outside, so it might be called an opaque house. An insulated cylinder (chimney) with a damper duct is set on the outside of the opaque house to make it fully passive function. Subsequently, the opaque house was carefully designed and actually constructed at the main site for the proving test; Ashoro (43°14.5′N, 143°33.5′E) and the sub-site for the proving test, Asahikawa (43°46′N, 142°22′E) in the eastern and northern parts of Hokkaido respectively. A proving performance test of the opaque house and the analysis of the results were successfully carried out by our project team for two and a half years. The capability of the solar lumber drying of the opaque house could be verified experimentally (Kanayama, et al., Aug., 2008), (Baba. et al., Oct., 2008), (Koga, et al., Oct., 2008).

2. A new concept of a solar lumber drying technology

2.1 Outline of the concept

An opaque house is covered by a composite surface consisting of a triple transparent film with double air layers and a CF-sheet, among which are a few spaces held by a skeleton frame. In this case, solar radiation (S. R.) incident upon the house surface is absorbed by the CF-sheet effectively, and converted into solar heat (infrared radiation) and only few part of the incidence is transmitted through the opaque house as solar ray ultimately. As shown in Fig.1, this optical and thermal mechanism is induced by a "transparent insulation/blackbody cavity effect" (Kanayama, K., et al., 2010), (Kanayama, K., et al., 2010). From this, an incident ray falls three-dimensionally into the cubical opaque house (=Volumetric S. R. incidence) and global S. R. from the sky is collected three-dimensionally by the opaque house (=Volumetric solar heat collected) too. In this case solar heat can be collected passively by the opaque house and saved passively into the opaque house. Hence, any electric power is needless. On the other hand, a moist air produced when drying lumber in the opaque house can be sucked by the draft force through an insulated cylinder set on the outside of the opaque house (Fig.5). This is a thermo siphon phenomenon caused by density difference due to temperature difference between inside air and outside air of the opaque house. As above, the combination of the composite surface consisting of a triple transparent film and a CF-sheet, and an insulated cylinder with a damper duct, could successfully collect S. R. Besides, the inside air is taken out from, and the outside air is taken into, the opaque house in accompany of the phenomenon. In this case, electric power is needless, so this is the reason for the name: "a fully passive-type solar lumber drying house".

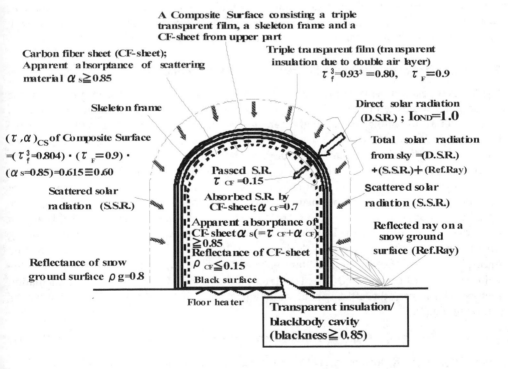

Fig. 1. "Transparent Insulation/Blackbody Cavity Effect" on the opaque house covered by a composite surface. (Kanayama, K. et al., Jun., 2009)

2.2 Explanation on the "transparent insulation/blackbody cavity effect"

Referring to Fig.1 and Fig.2, when a unit of S. R., I_{OND} (Normal direct radiation=1.0), was incident upon the opaque house covered by a composite surface, 0.804 of that is transmitted through triple transparent film, 0.85 of that is absorbed by the CF-sheet (=scattering medium with apparent absorptance 0.85), and 0.9 of that is passed through a skeleton frame, as discussed in the section 2.3 in detail. However, the CF-sheet inside the house absorbs 0.7, transmits 0.15, and the residual 0.15 is reflected toward the outside of the opaque house. Consequently, 0.615 ($\equiv$0.6) is trapped as solar heat (=infrared radiation) into the opaque house. Because the inside of the house is a closed space covered by the composite surface involving the CF-sheet, the solar heat under goes secondary reflection, absorption and re-emission as infrared ray in the cavity, so that a quasi-blackbody is realized or obtained with blackness of 0.85 or more, because that the transmitted ray of 0.15 is also incident on any body inside the house, and absorbed as heat.

A component transmitted outside from the inside of the house is very small due to triple transparent film, but mostly opaque for the infrared region. Heat (=near infrared and infrared radiation) of relatively high temperature inside the house is mostly, sometimes, shut out by the double air layers among the triple film, and due to a radiation property which is protective for infrared region, so that as a result, heat transfer loss is negligibly small. Thus, when S. R. is incident upon an opaque house it is mostly trapped within the opaque house as solar heat. Therefore, this phenomenon, because that the inside of the house is filled with infrared radiation beam, is called "transparent insulation/blackbody cavity effect".

2.3 Component materials and efficiency of volumetric collection

Fig.2 shows relation between radiation properties of the materials, consisting of composite surface, and intensity of direct S. R. incidence I_{OND} in detail. On the radiation property of each material, assuming that a single transparent film's transmittance is τ_f=0.93, a triple transparent film's transmittance is τ_f^3=0.804. If the CF-sheet's absorptance is α_S=0.85, and the passing rate of the skeleton frame is τ_F=0.90, so the coefficient of transmittance-absorptance of the composite surface becomes $(\tau \cdot \alpha)_{CS} = \tau_f^3 \cdot \alpha_s \cdot \tau_F = 0.804 \cdot 0.85 \cdot 0.90 = 0.614 \equiv 0.60$. Fig.3 shows angular relation (θ, a) of solar colleting surfaces A_n and a house model, consisting of multi-surface A_n, n=1....n. If the solar radiation (S. R.) incident on the multi-surface, made up of a number of composite surfaces, from A_1 to A_n, was known a intensity of $I(d)_{tilt}$, on a tilt surface, by multiplying the $I(d)_{tilt}$ into the corresponding surfaces, $A_1 \sim A_n$, and summing up each product, the volumetric S. R. incidence, $I(Q)_{VL}$ can be determined. Moreover, by multiplying coefficient of transmittance-absorptance $(\tau \cdot \alpha)_{CS}$ [=0. 6] into $I(Q)_{VL}$ the volumetric solar heat collected, Q_{VC}, is obtained. This is called as a conventional calculation method. Therefore, the efficiency of volumetric solar heat collected η_{VC} is defined as following Eq.(1):

$$\eta_{VC}=I(Q)_{VI}/I(Q)_{fl}\times(\tau \cdot \alpha)_{CS}=\eta_{VI}\times(\tau \cdot \alpha)_{CS} \tag{1}$$

where $I(Q)_{fl}$ is S. R. incidence on the floor, η_{VI} is the efficiency of the volumetric S. R. incidence. In this case, $I(d)_{tilt}$ in the database (NEDO's Report, (1997)) shown in Table 1 from AMeDAS is used as numerals at the experimental site; Ashoro (43°14.5'N, 143°33.5'E), Tokachi-pref., in Hokkaido.

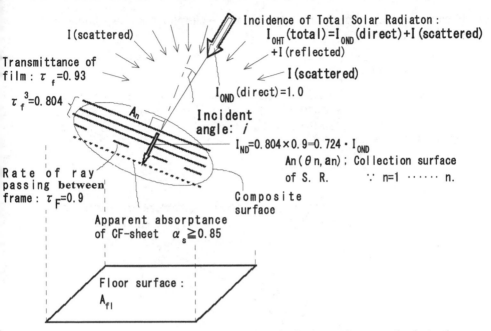

Fig. 2. Radioactive property of materials consisting the house and a mutual relation between incident S. R. and collecting surface

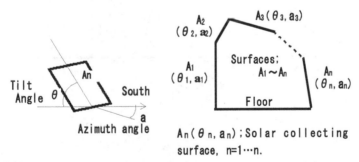

Fig. 3. Angle relation on a solar collecting surface A_n and a house model consisting of multi-surface A_n, n=1 · · · n

Surfaces	Angles of Tilt θ & Azimuth a	Jan	Feb	Mar	Apl	May	Jun	Jly	Aug	Spt	Oct	Nov	Dec	Year
Roof; I(d)rf=I(d)fl=I(d)HT =Horizontal Total S.R.	θ=0° a=0°	6.768	10.15	13.86	15.95	17.75	17.86	15.88	13.82	11.77	9.828	6.696	5.544	12.17
South Surface. ; I(d)ws	θ=90° a=0°	13.46	15.84	14.18	10.33	8.928	8.316	7.956	7.992	8.928	11.23	10.91	11.16	10.76
East-West Surfaces ; I(d)WE/WW	θ=90° a=90°	11.59	17.57	18.58	18.72	20458	19.94	17.64	15.41	14.04	12.89	9.360	8.748	15.41

Table 1. Solar Radiation (S. R.) Incidence on a Tilt Surfaces I(d)tilt at Experimental Site; Ashoro (43°14.5′N, 143°33.5′E) MJ/m²d (South-North model)

3. Calculation method of performance factors by a simplified model of an actual opaque house, (Kanayama, K., et al., Oct., 2008)

As shown in Fig.4, $(\tau \cdot \alpha)_{CS}$ corresponds to an efficiency of collection of volumetric S. R. incident upon the composite surface.

That is, $(\tau \cdot \alpha)_{CS}$ is always constant as 0.6 determined only by a radiation property of material, having no connection with S. R. intensity. In Fig.4, shows the efficiency, η_{AC}, of solar heat collection by a typical flat-type. However, η_{AC} is a function of Io_{ND}, changing in proportion to S. R. intensity.

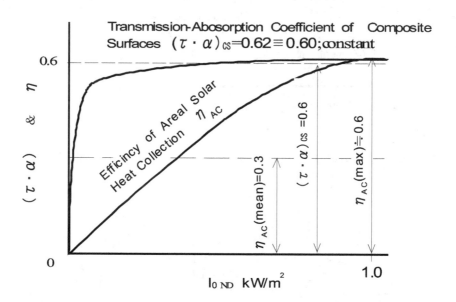

Fig. 4. Transmission-absorption coefficient for volumetric solar heat collection and an efficiency of areal solar heat colection

3.1 On the "South-North model"

On the database shown in Table 1 (NEDO's Report, 1997), $I(d)_{rf}$ is S. R. incident on a roof surface, $I(d)_{ws}$ is S. R. incident on a south wall, $I(d)_{we/ww}$ is S. R. incidence on an east and a west walls, all of which are tilt S. R. incidence per unit surface and per one day (MJ/m^2d) in each month respectively. Fig.5 shows a calculation model of an opaque house adopted in the conventional calculation method. The model of the opaque house (South-North model) is shaped a Quonset hut (floor area 5.0 m×5.0 m=25.0 m², height 3.4 m), besides two insulated cylinders which are set vertically outside the house. Thus, all the surfaces of the house is covered by a composite surface consisting of triple transparent film and CF-sheet, so that the actual house looks opaque. As volumetric capacity of the opaque house, can contain stacked lumber of net 10 m³ in maximum to be loaded along south-north direction.

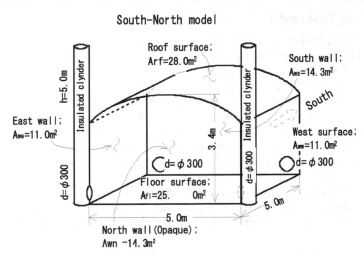

Fig. 5. A fully passive solar lumber drying house (South-North model)

Fig.6 shows an estimation, calculated by a conventional method, on the drying performance of the "South-North model" of the opaque house. During the season of spring to early summer, the volumetric solar heat collected Q_{VC} attained between 170~450 MJ/d. The rate of these numerals to S. R. incident on the floor is corresponds to an efficiency of volumetric solar heat collected η_{VC}, which are nearly to 120~200 %. η_{VC} is larger in winter and smaller in summer than yearly averaged value, thus its maximum value is 198 % in January and its yearly averaged value is 139 %.

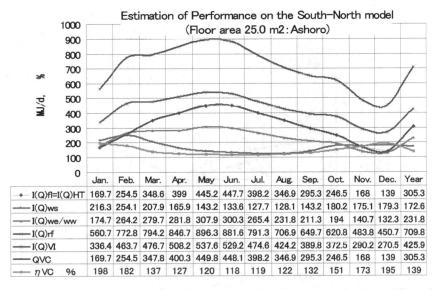

	Jan.	Feb.	Mar.	Apr.	May	Jun.	Jul.	Aug.	Sep.	Oct.	Nov.	Dec.	Year
I(Q)fl=I(Q)HT	169.7	254.5	348.6	399	445.2	447.7	398.2	346.9	295.3	246.5	168	139	305.3
I(Q)ws	216.3	254.1	207.9	165.9	143.2	133.6	127.7	128.1	143.2	180.2	175.1	179.3	172.6
I(Q)we/ww	174.7	264.2	279.7	281.8	307.9	300.3	265.4	231.8	211.3	194	140.7	132.3	231.8
I(Q)rf	560.7	772.8	794.2	846.7	896.3	881.6	791.3	706.9	649.7	620.8	483.8	450.7	709.8
I(Q)VI	336.4	463.7	476.7	508.2	537.6	529.2	474.6	424.2	389.8	372.5	290.2	270.5	425.9
QVC	169.7	254.5	347.8	400.3	449.8	448.1	398.2	346.9	295.3	246.5	168	139	305.3
η VC %	198	182	137	127	120	118	119	122	132	151	173	195	139

Fig. 6. Monthly and yearly estimated performance of the volumetric solar heat collected Q_{vc} and the efficiency of volumetric solar heat collection η_{vc}

3.2 On the "East-West model"

Table 2 shows a database from AMeDAS (NEDO's Report, 1997) at the proving test site. Fig. 7 shows a calculation scheme for the test for the "East-West model" of the opaque house. Similarly from Table 2, on the "South-North model", multiplying the daily S. R. on each tilted surface, per unit area, per day, i.e. $I(d)_{rf}$ on the roof surface, $I(d)_{fl}$ on the floor surface, $I(d)_{ws}$ on the south wall, and $I(d)_{we/ww}$ on east and west walls, by each area; A_{rf} of roof area, A_{fl} of floor area, A_{ws} of south wall, and $A_{we/ww}$ of east and west walls respectively, and by summing up all of them, the volumetric S. R. incidence $I(Q)_{vI}$ can be determined. Then multiplying the product of each term by the coefficient of transmittance-absorptance 0.6, and by summing up all the terms, thus the volumetric solar heat collected Q_{vc} can be determined. The results are shown in Fig. 8 with a data Table. According to the results of Fig 8, an efficiency of volumetric solar heat collection η_{vc} based on solar radiation incident on the floor area, 206 % is maximum in January, 117 % is minimum in June, and 141 % is average value for the year. These are the largest merit in the concept of volumetric solar heat collection.

Surfaces	Angles of Tilt θ & Azimuth a	Jan	Feb	Mar	Apl	May	Jun	Jly	Aug	Spt	Oct	Nov	Dec	Year
Roof S.; I(d)rf=I(d)fl =I(d)HT	θ=0° a=0°	6.768	10.15	13.86	15.95	17.75	17.86	15.88	13.82	11.77	9.828	6.696	5.544	12.17
Roof S.;I(d)rf=I(d)o'	θ=10° a=0°	8.665	12.17	15.61	16.96	18.32	17.92	16.23	14.40	12.72	11.33	8.300	7.202	13.34
South S. ; I (d)ws	θ=90° a=0°	13.46	15.84	14.18	10.33	8.928	8.316	7.956	7.992	8.928	11.23	10.91	11.16	10.76
East-West Ss.;I(d)we/ww	θ=90° a=90°	11.59	17.57	18.58	18.72	20.458	19.94	17.64	15.41	14.04	12.89	9.360	8.748	15.41

Table 2. Solar Radiation (S. R.) Incidence at Experimental Site; Ashoro (43°14.5'N, 143°33.5'E) MJ/m²d (East-West model)

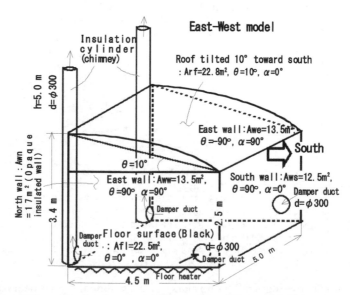

Fig. 7. A fully passive solar lumber drying house (East-West model)

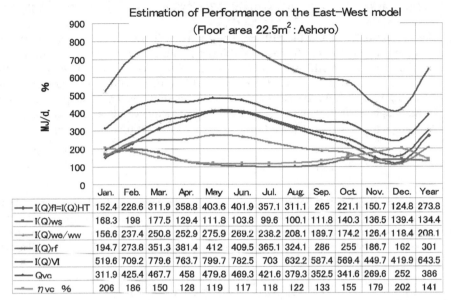

Fig. 8. Monthly and yearly estimated performance of the volumetric solar heat collected Q_{vc} and the efficiency of volumetric solar heat collection η_{vc}

4. Performance calculation and proving test of the opaque house

In the previous section, a performance analysis on "South-North model" and "East-West model" of an opaque house was carried out by applying numerals of the database of S. R. incidence and based on the performance factor of volumetric solar heat collected Q_{VC}, the results were discussed and compared each other. In this section, adopting the actual measurements of S. R. incidence at the proving test site, the performance analyses were carried out according to a normal calculation method as following, and then the results obtained were compared with the proving test (Baba, H., et al., 2008).

4.1 Volumetric solar heat collected by a normal calculation method

Applying the authors' separation method (Baba, H., Kanayama, K., July, 1985) for direct-scattered components of S. R. incidence, volumetric S. R. incidence on the opaque house can be calculated, and then the volumetric solar heat collected Q_{VC} can be obtained. This calculation process is fairly complicated, therefore, only an outline of the method will be given as follows: As shown in the left part of Fig.3, assuming tilt angle $\theta_n°$ and azimuth angle $a_n°$ on a tilt surface A_n respectively, the volumetric S. R. incidence $I(H)_{VI}$ upon an opaque house, can be calculated by multiplying S. R. incidence $I(\theta_n, a_n)$ by each tilt surface A_n of simplified model in the right part of Fig.3 and summing up all of each product $A_n \cdot I(\theta_n, a_n)$, n=1,2 $\cdot \cdot \cdot$ n. Therefore, using the same method, on the actual model of the opaque house, volumetric S. R. incidence upon the opaque house $I(H)_{VI}$ can be calculated by Eq.(2), and the volumetric S. R. incidence $I(Q)_{VI}$ and volumetric solar heat collected Q_{VC} of the actual opaque houses in Fig.5 and Fig.7, can be obtained.

$$I(H)_{VI} = A_{rf}I(\theta_{rf}, a_{rf}) + A_{we}I(90,-90) + A_{ws}I(90,0) + A_{ww}I(90,90) + A_{wn}I(90,-180) \qquad (2)$$

where, the first term, $\cdots\cdots$, and the fifth term of the right hand of Eq. (2), indicate the S. R. incident on the roof surface, on the east wall surface, on the south wall surface, on the west wall surface and on the north wall surface respectively. Where, $I(H)_{VI}$ is volumetric S. R. incidence in kJ/h, $I(\theta, a)$ is S. R. incidence on the tilt surface with setting angles (θ, a) in kJ/m²h. A_{rf} is roof area m², A_{we} is east wall area m², A_{ws} is south wall area m², A_{ww} is west wall area m², and A_{wn} is north wall area m². Now, the S. R. incidence $I(\theta, a)$ on a tilt surface A_{tilt} with setting angle (θ, a) can be calculated from Eq.(3) by substituting Eqs. (4)~(6) ·

$$I(\theta, a) = I_{ND}\cos i + I_{SC}\frac{(1+\cos\theta)}{2} + \rho_g I_{HT}\frac{(1-\cos\theta)}{2} \qquad (3)$$

where, I_{ND} is the direct S. R. incidence kJ/m²h, I_{SC} is the scattering S. R. incidence kJ/m²h, I_{HT} is the horizontal total S. R. incidence kJ/m²h, ρ_g is the reflectance of the earth, i is the incidence angle between incident ray and the normal to the tilt surface. By integrating Eq.(2) from sunrise time to sunset time, we obtained the volumetric S. R. incidence on a tilt surface par day.

$$\cos i = \sin h\cos\theta + \sin\theta\cosh\cos(A - a) \qquad (4)$$

$$\sin h = \sin\phi\sin\delta + \cos\phi\cos\delta\cos t \qquad (5)$$

$$\sin A = \cos\theta\sin t / \cosh \qquad (6)$$

where, h is the solar altitude angle °, A is the solar azimuth angle °, t is solar time in °, φ is the latitude of the site in °, δ is solar declination angle in°.

By the way, the normal calculation method was utilized to calculate the efficiency of volumetric solar heat collection η_{VC} using the real measurements of S. R. incidence at the proving test site; Ashoro, and the results obtained were plotted in Fig.13.

4.2 Solar radiation (S. R.) incidence on each surface, volumetric S. R. incidence and volumetric solar heat collected

A fully passive-type solar lumber drying house has the peculiar characteristics that not only S. R. incidence upon a roof surface, but also the S. R. incidence upon the wall surfaces around the opaque house can be utilized as solar heat collected. Fig.5 shows the calculation schemes of the "South-North model" of the opaque house. (floor area 25.0 m² (=5.0 m×5.0 m), height 3.4 m).

First, using Eq.(3) the S. R. incidence on the roof surface, and on each vertical wall, with every azimuth angle is calculated, and thus the volumetric S. R. incidence $I(H)_{VI}$ is determined by summing up the S. R. incidence upon each surface, and after substituting into Eq.(2) and integrating with the sun shine hour, volumetric S. R. incidence $I(Q)_{VI}$ overall the opaque house can be determined. The volumetric solar heat collected Q_{VC}, can be determined as the product of $I(Q)_{VI}$ and $(\tau \cdot a)_{CS}$ [=0.6] . Where, the roof surface of the Quonset hut is assumed to be a horizontal flat surface for simplification. However, Fig.6 shows the performance factors on the "South-North model" with data table calculated by the conventional calculation method, using the database of S. R. incidence at the proving test. Also in the same way Fig.7 shows the calculation scheme for the "East-West model", and also Fig.8 is the results on the "East-West model" with data table of the performance factors calculated by the conventional calculation method, using the database as above.

4.3 Photos and drawings of the opaque houses in detail

In order to compare two opaque houses being tested, the details of the relevant parameters, in depth in drawings are presented with Photo 1 and Photo 2, and in Figs.9 and 10. The actual scheme on the "South-North model" of the opaque house and its outside view when loading the stacked lumber on a trolley in Fig.9, and Photo 1, and that of the "East-West model" with its front view of the opaque house in Fig. 10 and Photo 2 respectively. Fig.9 and Fig.10 are the complete formation, dimensions and arrangements of the parts and so on, as shown by the drawings of the front picture and the plan picture in detail of the actual opaque houses, "South-North model" and "East-West model" respectively. In Fig.9, the "South-North model" has one set of two fan-convectors involving a fin-tube heater and a small fan (50 W×2) on the east floor of the house, and one set of four small fans (25 W×4) on the ceiling on the west side, and the same sort of parts and the same numbers as those are on the opposite floor and ceiling. Those two groups of each set are intermittently operated in the opposite direction in half a day in order to circulate uniformly the inside air as a breeze. In Fig.10, the "East-West model" has only one set of two fan-convectors (50 W×2 each) on the south floor of the house, and only one set of four small fans (25 W×4) on the ceiling part on the north side. Moreover, a floor heater (feed pump; 25 W×2) is molded in the black concrete floor, and we can see the stacked lumber (net volume 10 m³ in max.) in the central part of both houses. The small fan circulates slowly the inside air of the opaque house to increase the drying speed and to dry lumber uniformly. Hence, air speed of 0.2m/sec between the lumbers is sufficient due to the inside of the opaque house is filled with infrared radiation beam. While, the fan-cons and the floor heater contribute to heat supply for drying auxiliary, however, in summer season or in good weather conditions both auxiliary heats are not necessary.

Photo 1. A Fully Passive Solar Lumber Drying House Loading Larch Lumber (2x4 material); (South-North model)

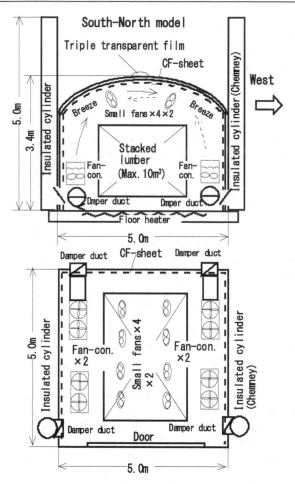

Fig. 9. A Fully Passive Solar Lumber Drying House (South-North model)

Photo 2. Front View of A Fully Passive-type Soalr Lumber Drying House; (East-West model)

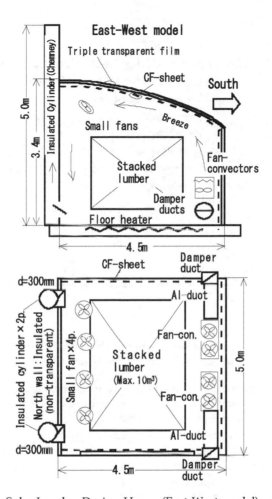

Fig. 10. A Fully Passive Solar Lumber Drying House (East-West model)

4.4 Hourly variation of S. R. incidence and solar heat collected, and the others

Fig. 11 shows hourly variations of volumetric S. R. incidence and volumetric solar heat collected over all the opaque house of "South-North model", after obtaining the hourly S. R. incidence on the tilt surface, such as a roof surface and every vertical wall surfaces, applying the normal calculation method to the actual S. R. measurements at the proving test site. In Fig.11, lumber drying test period of fourteen days, between Feb. 19th~Mar. 4th/' 07, in early spring, is a representative graph of several data; two day's graphs on Feb. 23rd (cloudy day) and 24th (fine day) are shown. From this Figure, we can see hourly variation of S. R. incidence on each surface, volumetric S. R. incidence $I(H)_{VI}$ kJ/h, volumetric solar heat collected q_{VC} kJ/h and air velocity in insulated cylinder V_{eX} 0.1m/s.

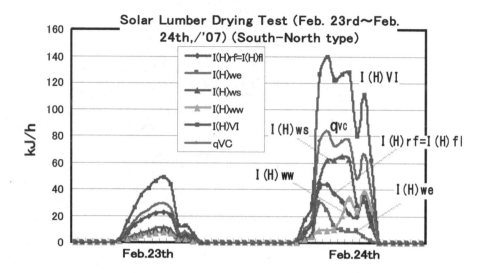

Fig. 11. Hourly variation of the measurement of volumetric S. R. incidence upon each surface and volumetric solar heat collected on cloudy day (23rd) and fine day (24th)

Fig.12 shows hourly variation of all the measured quantities inside and outside the "South-North model" on every day, during the fourteen days of the proving test. Where, T_o is outside temperature °C, T_i is inside temperature °C, H_o is outside humidity %, H_i is inside humidity %, $I(H)_{HT}$ is horizontal total S.R. incidence $kJ/m^2 h$, V_e is air velocity x0.1m/s in an insulated cylinder. Outside temperature changes between -20~0 °C every day, while the outside humidity changes inversely between 40~90 % every day. Inside temperature changes between 35~40 °C, with average value going slightly up near the end of drying period. Inside humidity is 40~45 % initially, however, with progress of lumber drying it gradually goes down to about 10 % near the end with varying wavy. Fig.13 shows the estimated efficiency of volumetric solar heat collected η_{VC} of "South-North model" and "East-West model" transferred from Fig.6 and Fig.8. The η_{VC} obtained by the normal calculation method, substituting the measured results of five proving tests during one year, from the first test in Oct. 31st~Nov. 15th/'06 to the fifth test in Aug. 16th~30th/'07 were plotted on the same figure for easy comparison of the quantities. The η_{VC} of "East-West model" is a little greater than that of "South-North model". Moreover, on the former a coincidence between the estimated value and the measured one can be seen. The yearly average estimated value of 140 % for "South-North model" is nearly equal to 141 % of "East-West model", but on the measured value of 157 % for the former, is always greater than 125 % of the latter. However, if several simplifications on the calculation process, and characteristics of the field test were considered, these numerical differences could be negligible.

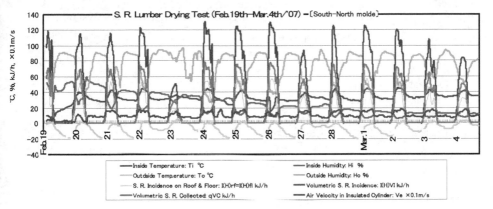

Fig. 12. Daily Variation on Main Data Inside and Outside the Solar Lumber Drying

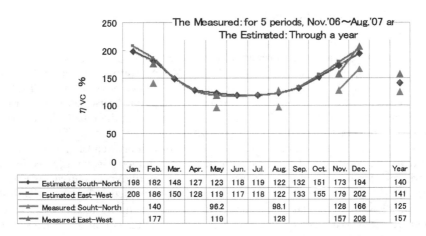

	Jan.	Feb.	Mar.	Apr.	May	Jun.	Jul.	Aug.	Sep.	Oct.	Nov.	Dec.	Year
Estimated: South-North	198	182	148	127	123	118	119	122	132	151	173	194	140
Estimated: East-West	208	186	150	128	119	117	118	122	133	155	179	202	141
Measured: Souht-North		140			96.2			98.1			128	166	125
Measured: East-West		177			110			128			157	208	157

Fig. 13. Comparison betweeen the Estimation and the Measurements of Efficiency of Volumetric Solar Heat Collected η_{vc}

Where, η_{vc} of the "East-West model" appears always to be greater than that of "South-North model", the reason is due to the fact that calculation assumed for the former that the roof surface is 10° tilted toward south as shown in Fig.7, while for the latter, that the surface is horizontal. Table 3 shows the averages of the measurements picked from Fig.10, and the other, from the initial two days and final two days. The difference of air density between the inside and outside of the opaque house, results in the difference of pressure between the inside and outside of the house. As a result, the negative pressure F_d in the insulated cylinder induced by Eq.(7) is a function of the difference of air density γ and the height of insulated cylinder h. The height of the insulated cylinder is 5.0 m and diameter is 300 mm.

$$F_d = h(\gamma_o - \gamma_i) \times 9.8 \ P_a \tag{7}$$

where, $(\gamma_o - \gamma_i)$ is the density difference between inside and outside of the house in kg/m³.

Items	South-North model	East-West model
Outside Temp. T_o / Outside Humid. H_o	-5.0 ℃ / 79.2 %	
Inside Temp. T_i / Inside Humid. H_i	36.7℃/13.0 %	37.0℃ /14.0 %
Total Heat Input Q_{in}	21.8 GJ	23.6 GJ
Efficiency of Volumetric Solar Heat Collection η_{vc}	166 %	208 %
Volumetric Solar Heat Collected Q_{vc}	3.6 GJ	4.3 GJ
Solar Heat Fraction F_s	16.4 %	18.3 %
Rate of Evaporation Heat R_{ev}	11.2 %	10.1 %
Water Evaporated W_{ev}	1010 kg	980 kg
Decreas of Moisture Contents ΔM_c	37%→9 % (D.B.)	38%→10 % (D.B.)

Table 3. Experimental Results on the 2nd Drying Test around a Winter Solstice (Dec. 13th~28th/'06)

The draft force is induced in the insulated cylinder as shown in Fig.12, air velocity in the insulated cylinder goes up over 1m/s when the inside temperature goes up in daytime, and goes down less than 1m/s when the inside temperature goes down at night. Table 3 also specially shows the important results in comparing the opaque houses with each other with their data measured around the winter solstice. The results of the fourth drying test (Feb. 19th~Mar. 4th/'07), in Fig.14, shows the performance factors of both "South-North" and "East-west" by a bar chart graph. S. R. incidence on a floor area of the opaque house, measured at the proving test site, the efficiency of volumetric solar heat collected η_{VC} estimated from the database in March was 140 % for "South-North model"; it was nearly equal to η_{VC}=148%. Thus, the differences in performance factors between "South-North model" and "East-West model" are caused by the fact that the former is a prediction estimated from the past statistics weather data and the latter is that given from a normal calculations using S. R. measured.

Similarly, based on S. R. incidence on the floor area of the opaque house, efficiency of volumetric solar heat collection η_{VI} is 232.4 % for "South-North model" and 294.8 % for "East-West model". The fact that "East-West model" 's efficiency is always larger than that of "South-North model" is caused by assuming that the roof surface of "East-West model" is tilted 10° toward south as described above. Fig.15 (Koga, S., et al. Oct., 2007) shows the result of drying test of larch lumber (2×4 material). Size of the specimen of lumber is 50 mm D×100 mm W×3,650 mm L and number of the specimen is 441 pieces, so that net volume of the stacked lumber is 8 m³. In general, net volume 10m³ of the lumber can be loaded in the opaque house. Larch lumber with initial moisture content 40 % (D.B.) was dried out less than 10% in two weeks, however, the other results of "South-North mode" and "East-West model" are about same with each other. In the case an auxiliary heat was supplied, the solar heat fractions were about same, 30 % in both cases. In summer, lumber drying by only solar heat is capable of drying out less than 20 % within two weeks, during which the solar heat fraction is 100 %. (Koga, S., et al., Oct., 2008)

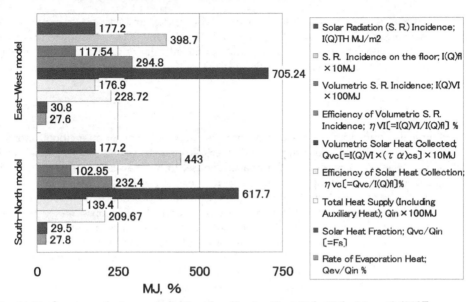

Fig. 14. Performance factor on Solar Lumber Drying Test; Feb. 19th~Mar. 4th (2007)

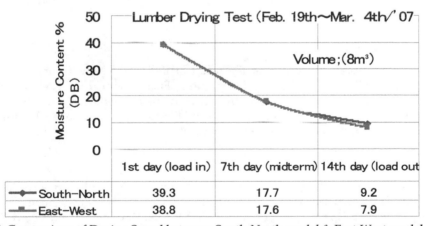

	1st day (load in)	7th day (midterm)	14th day (load out
South-North	39.3	17.7	9.2
East-West	38.8	17.6	7.9

Fig. 15. Comparison of Drying Speed between South-North model & East-West model

4.5 Progression on the working performance of the opaque house

The results of the research on the opaque house, as NEDO's project, carried out at the main site of the proving test; Ashoro, were described mostly. However, in the parallel activity, at the sub-site; Asahikawa, Hokkaido Forest Product Research Institute (HFPRI), which is a member of our project team, in order to grade up the performance of the opaque house, a patient examination has to be continued for more efficient usage, after the expiry of the national project. A technical advantage was found out for the laminar larch lumber, to produce high quality materials with less moisture distribution and less exude of resin than former (Kanayama, K., et al., Nov., 2010), (Koga, S., et al., Oct. 2008). Moreover, HFPRI is examining now how to dry not only lamina lumber but also large square sectional lumber with high quality as short as possible utilizing solar radiation only, without any auxiliary heat. Good result is expected.

5. Conclusions

First, the optical and the thermal mechanism of volumetric S. R. collection incident upon an opaque house covered by a composite surface were explained. Second, on a "South-North model" and a "East-West model" of the opaque house, under new concept of volumetric solar heat collection, the working performance of the opaque house was calculated applying S. R. database at the proving test site. As a result, one of the performance factors, η_{VC} of both houses, which was defined as efficiency of volumetric solar heat collection, attained 140 % of averaged value through the year. The rate of solar heat collected was 1.4 times of S. R. incident on the floor of the opaque house. Thirdly, applying a normal calculation method, using S. R. measured at the proving test site; Ashoro, a volumetric solar heat collected Q_{VC} of the opaque house constructed on the site was calculated, the proving test on larch lumber drying was carried out in cooperation with all staffs, and drying performance was verified to be good. Where, an electric power was supplied to feed pumps for auxiliary heat and fans to agitate inside air in the opaque house was 400 W maximum which was negligibly small. Thus, a "fully passive-type solar lumber drying house" proposed by us, could be achieved to dry out larch lumber under the conditions of low temperature and low humidity inside the opaque house.

6. Acknowledgements

This developing research is a part of the result of Project of "New Technology System Development Utilizing Solar Energy", by New Energy and Industrial Technology Department Organization (NEDO). On behalf of the authors, I would like to appreciate all the financial support, and at the same time, I would like to say thanks a lot to all staffs of the University, the Institute and the Company collaborated.

7. References

Baba, H., Kanayama, K., (July,1985) Studies of Measuring and Estimating of Solar Radiation (2nd Report), *Trans. JSME, B*, Vol.51, No.467, pp. 2451-2456, Tokyo, Japan.

Baba, H., Kanayama, K., Koga, S., Sugawara, T., (Oct.,2008) Development on a Fully Passive Solar Lumber Dying Apparatus (Part 2), *Journal of JSES*, Vol.34, No.5, pp. 57-64, ISSN0388-9564,Tokyo, Japan.

Baba, H., Kanayama, K., Koga, S., Sugawara, T. , (Jun.,2009) Performance Analysis and Proving Test on a Solar Lumber Drying House Covered by a composite Surface, *Proceedings of 46th NHTS of Japan*, pp.505-506, Kyoto, Japan, (CD-ROM) ISSN 1346-1532, G222.

Kanayama, K., Baba, H., (May, 2004) *Utilizing Technology of Solar Energy*, Morikita-Pubisher Ltd, Co., pp.212-214, ISBN 4 -627-94661-9, Tokyo, Japan.

Kanayama, K., Baba, H., Koga, S., Sugawara, T., (Oct., 2006) Developing Research on an Active-Passive type Solar Lumber Drying House, *Proceedings of Renewable Energy 2006*, pp.509-512, O-T-2-2 (CD-ROM), Chiba, Japan.

Kanayama, K., Baba, H., Koga, S., Seto, H., Sugawara, T., (Jul., 2007) Overview on Solar Lumber Drying Apparatus based on New Concept, *Proceedings of 2007 Symp. on Envir. Engg.*, pp.318-321, Osaka, Japan.

Kanayama, K., Baba, H., Koga, S., Sugawara, T., (Aug., 2008) Developing Research on a Fully Passive Solar Lumber Drying Apparatus (Part 1), *Journal of JSES*, Vol.34, No.4, pp. 39-47, ISSN 0388-9564,Tokyo, Japan.

Kanayama, K., Baba, H., Koga, S., Sugawara, T., (Oct., 2008) Developing Research on a Fully Passive Solar Lumber Drying Apparatus—Creation of a new concept and establishment of a soft technology— *Proceedings of Renewable Energy 2008*, O-ST-009, (CD-ROM), Busan, Korea.

Kanayama, K., Baba, H., Koga, S., Sugawara, T., (Jun., 2009) Solar Radiation Incidence and Thermal Analysis on an Opaque House Covered by a Composite Surface, *Proceedings of 46th NHTS of Japan*, pp.507-508, Kyoto, Japan, (CD ROM). ISSN1346-1532, G223.

Kanayama, K., Baba, H., Koga, S., Sugawara, T., (Jun.-July, 2010), Solar Heat Collected into an Opaque House Covered by Composite Surface in which Lumber Drying, *Proceedings of Renewable Energy 2010*, pp. 1-4, P-Th-1 (CD-ROM), Yokohama, Japan.

Kanayama, K., Baba, H., Koga, S., Sugawara, T., (Oct., 2010) Development Research on a Fully Passive Solar Lumber Drying Apparatus, —Creation of a new concept and establishment of a soft technology—*Current Applied Physics 10* (Oct., 2010), pp.s249-s253, ELSEVIER.

Kanayama, K., Baba, H., Koga, S., Tsuchihashi, H., Seto, H., Sugawara, T., (Nov., 2010), Characteristics on an Opaque Solar Lumber Drying House Covered by a Composite Surface, *Proceedings, of JSES/JWEA Joint Conference 2010*, pp.113-116, Koriyama, Japan.

Koga, S., Ohsaki, S.. Mabuchi, T., Ohgi, D., Cho, K., Yamauchi, K., Baba, H., Seto, H., Kanayama, K., Sugawara, T., (Oct., 2007) Developing Research on a Fully Passive-type Solar Lumber Drying House,—On the basis of experimental results from winter to summer—, (Oct., 2007) *Proceedings of JSES/JWEA Joint Conf. (2007)*, pp.185-188, Sapporo, Japan.

Koga, S., Kanayama, K., Baba, H., Sugawara, T., (Oct., 2008) Development on a Fully Passive Solar Lumber Drying Apparatus (Part 3), *Journal of JSES*, Vol.34, No.5, pp. 65-70, ISSN 0388-9564.

Miyoshi, A., (Sep., 1987) Development on a Fixed Heat process-type System, *Report of 7th Task Meeting of Branch*, New Energy Developing Organization (NEDO), pp.140-160, Japan.

NEDO's Report, (1997) Data-Map Relating to Solar Radiation around All the Country Based on the JWA's data, —Indication system of monthly averaged horizontal total S. R./S. R. on a tilt surface—, Ashoro, p.110, (CD-ROM,NP9703), or, http://www.nedo.go.jp/contents/MONSOLA00 (801), Ashoro, Japan.

Norota, T., Chiba, M., Nara, N., (Mar., 1983) Studies on Solar Drying of Lumber, *Report of Hokkaido Forest Products Research Institute*, No.72, pp. 97-125, CODEN:HRSKAC, Asahikawa, Japan,

Yamada, M., (May, 1998) "International Corroboration Research on Solar Energy Utilizing System (Indonesia)", *ENAA Report, Engineering No.79,* pp.16-19, Tokyo, Japan.

10

Development and Application of Asphalt Bonded Solar Thermogenerator in Small Scale Agroforestry Based Industry

R. S. Bello[1], S. O. Odey[2], K. A. Eke[1], M. A. Suleiman[3],
R. B. Balogun[1], O. Okelola[1] and T. A. Adegbulugbe[4]
[1]Federal College of Agriculture, Ishiagu
[2]Cross River University of Technology, Obubra, Cross River
[3]Federal College of Agriculture, Jalingo
[4]Federal College of Agriculture, Moore Plantation, Ibadan
Nigeria

1. Introduction

In today's climate of growing energy needs and increasing environmental concerns regarding energy shortages, scarcity and rapid depletion of non-renewable and environmental polluting energy resources such as fossil fuel, it is essential to diversify energy generation so as to conserve these fuels for premium applications hence the development, acceleration and use of new and renewable energy resources (Oladiran, 1999; Akarakiri and Ilori, 2003).

Recent concerns on the depletion of conventional energy sources such as agroforestry products and residuals such as wood and wood wastes have prompted interests in the use of solar energy for agricultural and forestry applications. Solar energy is an abundant and environmentally attractive alternative energy resource with enormous economic promises. In this era of energy shortages, it is noticed that the sun is an unfailing source of energy. It is free, the only disadvantage being the initial high cost of harnessing it. It is known that much of the world's required electrical energy can be supplied directly by solar power (Dennis and Kulsum, 1996).

The most commonly considered uses of solar energy are those classified as thermal processes. They include house heating, distillation of sea water to produce potable water, refrigeration and air conditioning, power production by solar-generated steam, cooking, water heating, and the use of solar furnaces to produce high temperatures for experimental studies (Encarta, 2002). Solar energy technologies such as photovoltaic cells, thermoelectric cell, thermionic cells, thermo emissive cells, etc are being used in small-scale applications on commercial projects (Encarta, 2002).

Electricity is also vital to modern day living without which there can be no meaningful development (Madueme, 2002). This is because in a technological and scientific development characteristic of the present day society, electricity is necessary for the operation of machines. The bulk of electrical power in the developing countries has been produced mainly from fossil-fuel based generating systems (Akarakiri and Ilori, 2003).

1.1 Solar energy applications

Specific areas of solar technology application has been identified to include electricity supply during power outages, telephone installations, industrial sector and drying of agricultural and forestry products like cocoa, timber etc (Akarakiri and Ilori, 2003). For instance, in Nigeria, Nitel powered the Ugonoba and the Gewadabawa repeater stations in 1997; more than 50 repeater stations in the Nigerian Network were powered by PV systems (Coker, 2004).

Solar energy has been limited mainly to low grade thermal applications in the Sub Saharan African region. For instance over 10,000 units of solar water heaters have been installed in Botswana, Zimbabwe, and South African (Akarakiri and Ilori, 2003). A project funded by the Agency for International Development (AID) in Tangaye, Africa provides fresh water and runs a grain mill for commercial production of flour (Maycock and Stirewait, 1981).

Several researches have been undertaken concerning the direct generation of electricity using the heat produced from nuclear reactors, kerosene lamps, firewood and biomass. The development of improved materials, use of multi-junction devices and novel cell designs to capture a higher proportion of the solar spectrum and use of concentration (Fresnel) lenses to focus the sunlight to high efficiency cells are areas of rapid development (Duffie, and Beckman, 1976).

The study of thermoelectric materials is a very active area of modern research that combines aspects of physical chemistry, solid state physics, and materials science. A thermoelectric material is a material that converts heat to electricity and vice versa. The main motivation for studying thermoelectricity is to find ways to improve their performance to better implement them in practical systems. The concept of thermoelectricity, a process that converts heat energy into electrical energy by using the Seebeck effect, has been used in agricultural operations.

1.2 Specific areas of thermoelectricity applications

Thermoelectric phenomenon has also been utilized for the accurate measurement of temperature by means of a thermocouple in which a junction of two dissimilar wires is maintained at a known reference temperature (for example, in an ice bath) and the other junction at the location where the temperature is to be measured. At moderate temperatures (up to about 260°C /500°F), wire combinations of iron and copper, iron and constantan (a copper-nickel alloy), and copper and constantan are frequently used. At high temperatures (up to 1649°C/3000° F), wires made from platinum and a platinum-rhodium alloy are employed (Encycl. Britannica, 1987).

Lertsatitthanakorn (2007) investigated the feasibility of adding a commercial TE module made of bismuth-telluride based materials to the stove's side-wall, thereby creating a TE generator system that utilizes a proportion of the stove's waste heat. The results showed that the system generates approximately 2.4 W when the temperature difference is 150 °C. This generated power is enough to run a small radio or a low power incandescent light bulb. Other research works (Rowe, 2006, Lertsatitthanakorn, 2007) were conducted in order to investigate the feasibility of using a TE generator in an improved biomass fired stove already developed.

The main motivation for studying thermoelectrics is to find ways to improve their performance for better implementation in agricultural systems (Champier et al., 2009). In areas with unreliable electricity supply, the feasibility of adding a commercial thermoelectric (TE) module to stove design is being investigated (Rowe, 2006). Nuwayhid et al., (2003) considered the prospect of applying TE modules in rural domestic woodstoves in regions where the electric supply is unreliable and subject to frequent disruption. This research work investigated the potentials of the utilization of asphalt bonded thermocouples in electricity generation and its application in small scale agricultural lighting projects such as poultry house illumination. The generator design work was made using existing high-quality Peltier modules in the power-generating mode.

1.3 Project objective

Demands for energy in agricultural production processes, especially in poultry production processes and other agro-forestry processes, are growing worldwide. In animal husbandry, light is an important aspect of the animal's environment. Avian species as well as mammalian species respond to light energy in a variety of ways for growth and reproductive performance. Acceptable system economic performances through extension services (Okelola et al., 2011) and system analysis (Nuwayhid et al., 2005) have been demonstrated at various levels. Despite the growing worldwide demand for energy in agricultural production processes, there has been an increase in production costs coupled with depletion of nonrenewable energy resources; therefore alternative and more attractive solar energy sources becomes imperative.

The concept of thermoelectricity using asphalt heating was employed in the development of TEG for lighting in agricultural projects such as small scale poultry house. The aim is to investigate the potentials of thermocouples embedded in asphalt (heat absorber) for electricity generation and application in agricultural projects.

1.4 Project justification

Application of solar energy has speedily improved the present low level of electricity production for domestic lighting and agricultural operations in several developing nations. However, according to the United Nations Development programme, 400milion families (nearly two billion people) have no access to electricity to light their homes, among other services (SELF, 2001).

The power generation potential of several energy harvesting modalities have been investigated (Roundy, 2003). While a wide variety of harvesting modalities are now feasible, solar energy harvesting through photo-voltaic conversion provides the highest power density, which makes it the modality of choice to power an embedded system that consumes several megawatt of energy using a reasonably small harvesting module. However, the design of a solar energy harvesting module involves complex tradeoffs due to the interaction of several factors such as the characteristics of the solar cells, chemistry and capacity of the batteries used (if any), power supply requirements, application behaviour, and power management features of the embedded system etc. It is, therefore, essential to thoroughly understand and judiciously explore these factors in order to maximize the energy efficiency of a solar harvesting module.

1.5 Energy storage technologies

The two choices available for energy storage are batteries and electrochemical double layer capacitors, known as ultracapacitors. Batteries are a relatively mature technology and have a higher energy density (more capacity for a given volume/weight) than ultracapacitors, but ultracapacitors have a higher power density than batteries and have traditionally been used to handle short duration power surges.

Recently, such capacitors have been explored for energy storage, since they are more efficient than batteries and offer higher lifetime in terms of charge-discharge cycles. However, they involve leakage (intrinsic and due to parasitic paths in the external circuitry), which precludes their use for long-term energy storage. While it is also possible to implement energy storage mechanism using an ultracapacitor and a battery, it is a tradeoff to a decrease in harvesting circuit efficiency due to the increased overhead cost of energy storage management.

2. Solar generator

Two components are required to have a functional solar energy generator; they are the collector and a storage unit. The collector simply collects the radiation that falls on it and converts a fraction of it to other forms of energy; either electricity and heat or heat alone. Methods of collecting and storing solar energy vary depending on the uses planned for the solar generator. The storage unit is required because of the non-constant nature of solar energy; at certain times only a very small amount of radiation will be received.

At night or during heavy cloud cover, for example, the amount of energy produced by the collector will be quite small. The storage unit can hold the excess energy produced during the periods of maximum availability, and release it when the productivity drops. In practice, a backup power supply is usually added, too, for the situations when the amount of energy required is greater than both what is being produced and what is stored in the container.

2.1 Thermoelectric Generator (TEG)

The conversion of sunlight into electrical energy in a solar cell involves three major processes; 1) absorption of the sunlight by solar cell (heat source) at a temperature T_H; 2) heating up of the thermocouple junction thus obtaining temperature difference between the ends of metal wires and thermoelectric potentials developed along the wire; and 3) the transfer of these separate thermoelectric potentials, in the form of electric current, to an external system.

A thermoelectric generator is a device that converts heat energy directly into electrical energy using Seebeck effect. This requires a heat source, a thermocouple and reference material. Thermoelectric generator is composed of at least two dissimilar materials, one junction of which is in contact with a heat source and the other junction of which is in contact with a heat sink. The power converted from heat to electricity is dependent upon the materials used, the temperatures of the heat source and sink, the electrical and thermal design of the thermocouple, and the load of the thermocouple (Angrist, 1982). Although TEGs have very low efficiencies (5 to 10 % in the above mentioned applications), their usage makes sense where the heat source is freely available and would otherwise be lost to the environment (Richner et al, 2011).

2.2 The principles of thermoelectricity

When two dissimilar metals are connected (i.e. welded or soldered together) to form two junctions, the voltage generated by the loop is a function of difference in temperatures between the two junctions. Such loop is called a thermocouple and the emf generated is called a thermoelectric emf. This thermoelectric phenomenon known as 'Seebeck Effect' was discovered in 1821 by Thomas J. Seebeck (Maycock et al., 1981). If this circuit is broken at the center, the net open voltage (the Seebeck voltage) is a function of the junction temperature and the composition of the two metals.

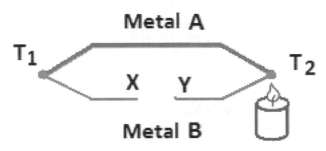

Fig. 1. A thermocouple made out of two different materials

Thermoelectric effects (Seebeck, Peltier, and Thomson)

There are three main thermoelectric effects experienced in thermo generation: the Seebeck, Peltier, and Thomson effects (Goldsmid, 1995; Nolas *et al.*, 2001). Thermoelectric effects are described in terms of three coefficients: absolute thermoelectric power (S), the Peltier coefficient (Π) and the Thomson coefficient (τ), each of which is defined for a homogenous conductor at constant temperature. Thermoelectric effects have significant applications in both science and technology and show promise of more importance in the future. Practical applications of thermoelectric effects include temperature measurement, power generation, cooling and heating (Steven, 2010).

Seebeck Effect: The Seebeck effect is responsible for the operation of a thermocouple. If a temperature gradient is applied across a junction between two materials, a voltage will develop across the junction, with the voltage related to the temperature gradient by the Seebeck coefficient.

For instance, the Seebeck coefficient of the thermocouple shown in Figure 1 as given by Steven, (2010) is:

$$\propto_{AB} = \frac{dV}{dT} \tag{1}$$

Where V is the voltage between points x and y, in the case of the thermocouple shown in Figure 1, $\Delta T = T_2 - T_1$ Then,

$$v_{xy} = \propto_{AB} (T_2 - T_1) \tag{2}$$

If T_2 is fixed and α_{AB} is known, then by measuring V_{xy}, one can determine T_1. If this circuit is broken at the center, the net open voltage (the Seebeck voltage) is a function of the junction temperature and the composition of the two metals.

Peltier Effect: The Peltier effect is the opposite of the Seebeck effect in which electrical energy is converted into thermal energy. It is observed in applications such as thermoelectric coolers. The Peltier effect is an effect whereby heat is liberated or absorbed at the junction between two materials in which a current I is flowing through. The heat liberated (or absorbed) at the junction is given by

$$Q = \Pi_{AB} I \tag{3}$$

Where Π_{AB} is the Peltier coefficient (Nolas *et al.*, 2001),I is current)

Thomson Effect: If there is a temperature gradient across a material and a current is flowing through the material, then heat will be liberated or absorbed. This is Thomson effect, and it is described by the following equation:

$$Q = \tau I \frac{dT}{dx} \tag{4}$$

Where τ is the Thomson coefficient (Nolas *et al.*, 2001). The following relationships hold for two materials A and B (Nolas *et al.*, 2001):

$$\tau_A - \tau_B = T \frac{d\alpha_{AB}}{dT} \tag{5}$$

$$\Pi_{AB} = \alpha_{AB} T \tag{6}$$

The coefficients α_{AB} and Π_{AB} defined above are for a system consisting of two different materials with a junction between them. However, it is often useful to define absolute thermoelectric coefficients which describe a single material. All of the above equations are generalized when a single material is being discussed (Nolas *et al.*, 2001). If a material is subjected to a temperature gradient ΔT, then $\Delta V = \alpha \Delta T$, where α is the Seebeck coefficient and ΔV is the volumetric change of the material respectively. Relations (5) and (6) also apply to absolute thermoelectric coefficients, i.e.

$$\tau = T \frac{d\alpha}{dT} \tag{7}$$

and

$$\Pi = \alpha T \tag{8}$$

2.3 Measuring Seebeck voltage

The Seebeck voltage cannot be measured directly because a voltmeter must first be connected to the thermocouple, and the voltmeter leads create a new thermoelectric circuit. Connecting a voltmeter across a copper-constantan (Type T) thermocouple and observing the voltage output: the voltmeter should read only V_1, but by connecting the voltmeter in an attempt to measure the output of Junction J_1, two more metallic junctions have been created: J_2 and J_3 (Figure 2a).

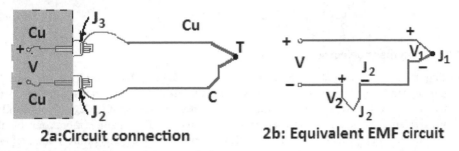

2a:Circuit connection   **2b: Equivalent EMF circuit**

Fig. 2. Measuring Seebeck voltage

Since J_3 is a copper-to-copper junction, it creates no thermal EMF ($V_3 = 0$), but J_2 is a copper-to-constantan junction which will add an EMF (V_2) in opposition to V_1 (Figure 2b). The resultant voltmeter reading V will be proportional to the temperature difference between J_1 and J_2. This says that we cannot find the temperature at J_1 unless we first find the temperature of J_2.

In a closed circuit composed of two linear conductors of different metals, a magnetic needle would be deflected if, and only if, the two junctions were at different temperatures giving rise to an emf being generated. The magnitude of the emf generated by a thermocouple is measured on standard by an arrangement shown in figure below.

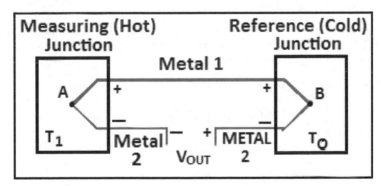

Fig. 3. Voltage measurement arrangement between two junctions

Points A and B are called junctions. Each junction is maintained at a well-controlled temperature (T_o or T_1) by immersion in a bath (cold junction) or connected to a heat source (hot junction). From each junction, a conductor (metal 2) is connected to a measuring device (potentiometer or an amplifier of low input impedance), which measures the thermal emf.

The Seebeck emf is approximately related to the absolute junction temperatures T_0 and T_1 by

$$V = \propto (T_1 - T_o) + \gamma(T_1^2 - T_o^2) \qquad (9)$$

Where α and γ are constants for the thermocouple pair.

The differential coefficient of equation above is the sensitivity, or thermoelectric power S, of the thermocouple

$$S = \frac{dV}{dT_1} = \propto +2\gamma T_1 \; (\mu V/\; 0 \, c) \tag{10}$$

2.4 Measuring thermoelectric power, S

According to the experimentally established law of Magnus, the thermoelectric emf for homogenous conductors depends only on the temperatures of the junctions and not on the shapes of the samples. This emf can thus be described by the symbol E_{AB} (T_0, T_1) and is determined solely by conductor A and can be written as E_A (T_0, T_1). This emf is more conveniently expressed in terms of a property, which depends upon only a single temperature. Such property is the absolute thermoelectric power (thermo power S_a (T)) defined as in equation (11).

$$E_A(T_0, T_1) = \int_{T_0}^{T_1} S_A \, (T) dt \tag{11}$$

If E_A (T_1, $T+\Delta T$) is known, for example, from measurements involving a super conductor, then S_A (T) can be determined from equation 12

$$S_A(T) = \lim_{\Delta T \to 0} E_A \frac{(T_1, T+\Delta T)}{\Delta T} \tag{12}$$

If equation 11 is for any homogenous conductor, then it ought to apply to both sides of the thermocouple. Indeed, it has been verified experimentally that the emf E_{AB} (T_0, T_1) produced by a thermocouple is just the difference between the emfs calculated using equation (17), produced by its two arms as follows.

Employing the usual sign correction, to calculate $E_{AB}(T_0, T_1)$, begin at the cooler bath, integrate $S_A(T)dT$ along conductor (metal 2) at junction A up to the warmer bath and then return to cooler bath along conductor (metal 1) through junction B by integrating $S_B(T)dT$.

This circular loop produces E_{AB} (T_0, T_1) given by equation 13 and 14.

$$E_{AB}(T_0, T_1) = \int_{T_0}^{T_1} S_A \, (T) dt + \int_{T_0}^{T_1} S_B \, (T) dt \tag{13}$$

$$E_{BA}(T_0, T_1) = \int_{T_0}^{T_1} S_A \, (T) dt - \int_{T_0}^{T_1} S_B \, (T) dt \tag{14}$$

Alternatively, combining the two integrals in equation 14

$$E_{AB}(T_0, T_1) = E_A(T_0, T_1) - E_B(T_0, T_1) \tag{15}$$

$$E_{AB}(T_0, T_1) = \int_{T_0}^{T_1} [S_A \, (T) dt - S_B(T)] \, dt \tag{16}$$

According to equation 16, $E_{AB}(T_0, T_1)$ can be calculated for a given thermocouple whenever the thermo powers $S_A(T)$ and $S_B(T)$ are known for the two constituents over the temperature range T_0 to T_1.

Defining S_{AB} according to equation 16 then yields equation 17

$$E_{AB}(T_0, T_1) = \int_{T_0}^{T_1} S_{AB} \quad (T) dT \tag{17}$$

$$\Rightarrow S_{AB}(T) = S_A \, (T) - S_B(T) \tag{18}$$

These equations lead directly to experimentally and theoretically verified results that in a circuit kept at a uniform temperature (dt = 0) throughout, E= 0 even though the circuit may consist of a number of different conductors (equation 18). If E did not equal 0, the circuit could drive an electric motor and make it perform work. The only source of energy would be heat from the surrounding. A circuit composed of single, homogenous conductor cannot produce thermoelectric emf (equations 6) when S_B (T) is equal to S_A (T). Finally, equation 6 makes it clear that the source of the thermoelectric emf in a thermocouple lies in the bodies of the two materials of which it is composed rather than the junctions.

The reference junction

One way to determine the temperature of reference junction J_2 is to physically put the junction into an ice bath, forcing its temperature to be 0°C. Since the voltmeter terminal junctions are now copper-copper, they create no thermal emf and the reading V on the voltmeter is proportional to the temperature difference between J_1 and J_2.

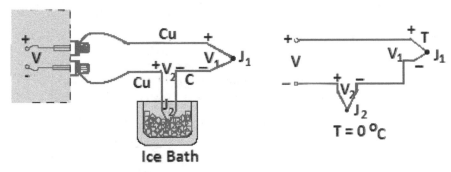

Fig. 4. External reference junction

Now the voltmeter reading is (see Figure 4):

$$V = V_1 - V_2 \cong \propto (t_{j1} - t_{j2}) \qquad (19)$$

If we specify T_{J1} in degrees Celsius:

$$T_{j1}(°C) + 273.15 = t_{j1} \qquad (20)$$

Then V becomes:

$$V = V_1 - V_2 = \propto [(T_{j1} + 273.15) - T_{j2} + 273.15)] \qquad (21)$$

$$V = \propto (T_{j1} - 0) = \propto T_{j1} \qquad (22)$$

This derivation is used to emphasize that the ice bath junction output, V_2, is not zero volts. It is a function of absolute temperature. By adding the voltage of the ice point, reference junction, we have now referenced the reading V to 0°C. This method is very accurate because the ice point temperature can be precisely controlled. The ice point is used by the National Bureau of Standards (NBS) as the fundamental reference point for their thermocouple tables, so we can now look at the NBS tables and directly convert from voltage V to Temperature T_{J1}.

Using an iron-constantan (Type J) thermocouple instead of the copper-constantan, the iron wire (Figure 5) increases the number of dissimilar metal junctions in the circuit, as both voltmeter terminals become Cu-Fe thermocouple junctions.

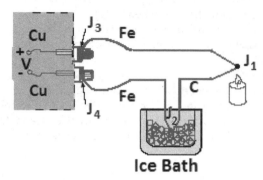

Fig. 5. An Iron- constantan couple

If both front panel terminals are not at the same temperature, there will be an error. For a more precise measurement, the copper voltmeter leads should be extended so that the copper-to- iron junctions (J_3 and J_4) are made on an isothermal (same temperature) block.

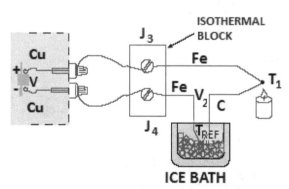

Fig. 6. Removing junctions from DVM terminals

The isothermal block is an electrical insulator but a good heat conductor, and it serves to hold junctions J_3 and J_4 at the same temperature. The absolute block temperature is unimportant because the two Cu-Fe junctions act in opposition.

Reference circuit

Replacing the ice bath with another isothermal block (Fig. 7) at J_{REF}, the new block is at reference temperature T_{REF}, and because J_3 and J_4 are still at the same temperature, we can again show that

$$V = \propto (T_1 - T_{REF}) \qquad (23)$$

A thermistor, whose resistance R_T is a function of temperature, provides us with a way to measure the absolute temperature of the reference junction. Due to the design of the isothermal block, junctions J_3 and J_4 and the thermistor are all assumed to be at the same temperature.

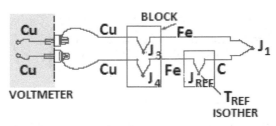

Fig. 7. Eliminating the ice bath

However, the use of thermocouples other than thermistor is more preferable in that thermocouple can be used over a range of temperatures, and optimized for various atmospheres. Thermocouples are much more rugged than thermistors, as evidenced by the fact that thermocouples are often welded to a metal part or clamped under a screw. They can be manufactured on the spot, either by soldering or welding. In short, thermocouples are the most versatile temperature transducers available and, since the measurement system performs the entire task of reference compensation and software voltage to-temperature conversion, using a thermocouple becomes as easy as connecting a pair of wires.

2.5 Thermogenerator composite parameters

The effective composite materials for thermoelectric generator include heat source, the thermocouple, the heat sink and the load. The composite parameters of the generator are the total Seebeck coefficient of the junction S, the total internal resistance r, and the total thermal conductivity k. The Seebeck coefficient can have either a positive or a negative sign. A material that has a negative S is referred to as an n-type material while a material having positive S is referred to as a p-type material.

Assuming that all the composite parameters are independent of temperature, and the Seebeck coefficients of the legs of thermocouple are S_n and S_p, and the electrical resistivity of each leg is ρ_n and ρ_p, and the thermal conductivities of each leg is k_n and k_p, then the parameters are defined by the following equations.

$$S = S_p - S_n = |S_p| + |S_n| \qquad (24)$$

$$r = \frac{\rho_n l_n}{A_n} + \frac{\rho_p l_p}{A_p} \qquad (25)$$

$$k = \frac{k_n A_n}{l_n} + \frac{\rho_p A_p}{l_p} \qquad (26)$$

Where l_n, A_n, and l_p, A_p, refers to the length and area of the n-type and p-type materials respectively, the heat source has a temperature T_n, and the heat sink has temperature T_c.

2.6 Thermogenerator efficiency

The efficiency η of generator is the power output I^2R divided by the heat input Q_{in}.

$$\eta = P_{out}/Q_{in} = I^2R/Q_{in} \tag{27}$$

R is the electrical load resistance. Heat input Q_{in} consists of the Peltier heat ST_nI plus the conduction heat $K(T_n-T_c)$ less one-half of the Joule heat I^2r librated in the thermocouple legs, i.e.

$$Q_{in} = ST_nl + k(T_n - T_c) - \frac{I^2r}{2} \tag{28}$$

Losses in maintaining the temperature T_n are not considered by this efficiency and thus it is not a total efficiency including heat source losses.

The ratio of load resistance R to internal resistance r is defined as $m = R/r$

The open-circuit voltage,

$$V_o = S(T_n - T_c) \tag{29}$$

The current,

$$I = \frac{V_o}{R+r} \tag{30}$$

Using these quantities, and selecting m to give optimum loading, the optimized efficiency, the efficiency expression becomes

$$\eta = \frac{T_n-T_c}{T_n}\left\{\frac{M-1}{M+{}^{T_c}/_{T_n}}\right\} \tag{31}$$

Where

$$M = m|_{d_n/d_{m=0}} = \sqrt{Z}(T_h - T_2)/2) \quad M = m|_{d_n/d_{m=0}} = \sqrt{Z}(T_h - T_2)/2) \tag{32}$$

Z is the figure of merit.

This efficiency for an optimum load consists of a Carnot efficiency η_c and device efficiency η_d thus

$$\eta_c = \frac{T_n-T_c}{T_n} \tag{33}$$

$$\eta_d = \frac{M-1}{M+{}^{T_c}/_{T_n}} \tag{34}$$

The device efficiency η_d will be a maximum for the largest value of M, for a fixed T_n and T_c; this requires a maximum value of Z. For most good thermoelectrics, $Z(T_n+T_c)/2 \approx 1$, so for $T_n/T_c \approx 1$, the efficiency is about 20% of the thermodynamic limit.

2.7 Figure of merit

This is a measure of the ability of a given thermoelectric material in power generation, heating or cooling at a given temperature T. ZT is given by the equation.

$$ZT = S^2 \sigma T / k \tag{35}$$

Where
S = the thermo power of the material
σ = The electrical conductivity
K = The thermal conductivity

The largest values of ZT are attained in semimetals and highly doped semiconductors, which are the materials normally used in practical thermoelectric devices. Figure of merit for single materials and thermocouples formed from two such materials varies hence one thermocouple can be better than another at one temperature but less effective at a second temperature

Z depends upon the material parameters S_p, K_p, ρ_p S_n, K_n, ρ_n and the dimensions of the two legs A_p, ℓ_n, ℓ_p, A_n . Maximizing Z with respect to the area-to-length ratio of the legs gives

$$Z|_{d_z/d_{x=0}} = \frac{(S_p - S_n)^2}{[(k_p l_p)^{1/2} + (k_n l_n)^{1/2}]^2} \tag{36}$$

When equation

$$\left[\frac{k_n l_n}{k_p l_p}\right]^{1/2} = \left[\frac{A_p l_n}{A_n l_p}\right] \tag{37}$$

For the optimum area-to-length ratio Z depends only upon the specific properties of the thermoelectric material. Generally, the parameters S, K, and ρ are not independent of temperature, and in fact the temperature dependence of the n and p legs may differ radically. The most widely used generator materials are lead telluride, which has a maximum figure of merit of approximately 1.5×10^{-3} K^{-1}. It can be doped to produce both p-type and n-type material and has a useful temperature range of about 300-700K (80-800°F). In material development, existing thermoelectric p and n materials operates from 300 to 1300K (80 to 1900°F) and yield an overall theoretical thermal efficiency of 18%.

To maximize power output, it is necessary to produce the largest possible voltage, thus Seebeck coefficient S should be made large, and hence proper selection of materials are required. Materials should have low electrical resistance in the generator. The legs should also have low thermal conductivities K since heat energy is carried away by thermal conduction. Hence the requirements for materials to be used in thermoelectric power generators are high S, low ρ and K and high figures of merit Z. Since the figures of merit Z for single materials vary with temperature, so do the figures of merit for thermocouples formed from two materials.

3. The thermocouple system

Thermocouples are differential temperature-measurement devices. They are constructed with two wires of dissimilar metals. One wire is pre-designated as the positive side (Copper,

Iron, Chromel) and the other as the negative (Constantan, Alumel). Basic system suitable for the application of thermoelectricity in power generation is that of several thermocouples connected in series to form a thermopile (a device with increased output relative to a single thermocouple). The junctions forming one end of the thermocouple are at the same low temperature T_L and the other junctions at the hot temperature T_H.

The thermopile is connected to a device in which the temperature T_L is fixed when connected to a heat sink. The temperature T_H is determined by the output of the heat source and the thermal output of the thermopile. The load is run by the charges generated. With a thermopile, the multiplication of thermocouple involves a corresponding increase of resistance, hence it follows that one thermocouple can be better than another at one temperature but less effective at a second temperature. In order to take maximum advantage of the different materials, the thermocouples are cascaded, producing power in stages and increase power output.

3.1 The choice of thermocouple

A primary consideration in choosing which thermocouple type to use in a given circumstance is the range of temperatures over which the device is to be used. Some of the other selection factors among others to be addressed include:

- Suitability for conditions of use, expected service life and compactable installation requirements
- Adequate sensitivity S over a wide range of temperature, stability against physical and chemical changes under various conditions of use and over extended periods of times,
- Availability, moderate costs, abrasion and vibration resistance.

Thermocouples can either be sheathed or beaded with bare thermoelements (Figure 8).

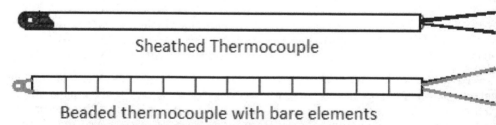

Fig. 8. Thermocouple materials

Sheathed thermocouple probes are available with one of the three junction types:

Grounded Junction Type: This is recommended for gas and liquid temperatures and for high-pressure applications. It has faster response than the ungrounded junction type.

Ungrounded Junction Type: This is recommended for measurements in corrosive environments where it is desirable to have the thermocouple electronically isolated from and shielded by the sheath.

The Exposed Junction Type: This is recommended for the measurement of static or flowing non-corrosive gas temperature where fast response time is required.

ANSI Polarity: In the thermocouple industry, standard practice is to colour the negative lead red. The negative lead of a bare wire thermocouple is approximately 6mm ('4'') shorter than the positive lead and the large pin on a thermocouple connector is always the negative conductor (Omega Eng., 2001) Standard Diameters of thermocouple available are: 0.25mm (0.010''), 0.50mm (0.020''), 0.75mm (0.032''), 1.0mm (0.04''), 1.5mm (1/16''), 3mm (1/8''), 4.5mm (3/16''), 6 mm (1/4''). With two wires 8mm and 9.5mm standard Omega thermocouples have 12-inch (300mm) immersion lengths. Other lengths are available.

3.2 Standard thermocouple types

ASTM and ANST standards explicitly stated that the letter designations identifying only the reference tables might be applied to any thermocouple with a temperature-emf relationship that complies with the table within the specified tolerances, regardless of the chemical composition of the thermocouple (Quinn, 1983). Any randomly chosen pair of dissimilar wires will produce some kind of thermal emf when subjected to a temperature difference from end to end, however, the emf so produced may be unpredictable and of little use. Hence certain thermoelement combinations have been commercially developed over the years that have proved to be useful, reproducible, and readily available.

Eight of the most widely used of these combinations have been assigned letter-designations for ease of reference, Base metal thermocouples types designated as E, J, K, and T. The rear (Noble) metal thermocouple types are S, R, and B types.

3.3 Base metal types

Type E, Chromel (nickel-10% chromium) (+) vs. Constantan (nickel-45% copper (-)). Type E is recommended for use to 900°C (1600°F) in oxidizing or inert atmospheres. Type E has been recommended as the most suitable of all standardized types for general low-temperature use, about -230°C (-380°F), since it offers the best overall combination of desirable properties i.e. high thermopower, low thermal conductivity, and reasonably good thermoelectric homogeneity typical values for the thermopower of type E at 4, 20, and 50K are 2.0, 8.5 and 18.7μVK^{-1} respectively (Spark et al., 1972).

Type J, Iron (+) vs. Constantan (nickel-45% copper (-)) is one of the most commonly used thermocouples in industrial pyrometry due to its relatively high thermopower and low cost. These thermocouples are suitable for use in vacuum, air, reducing, or oxidizing atmospheres to 760°C (1400°F) in the heavier gage sizes. Rapid oxidation of the iron wire at temperatures above 540°C (1000°F) limits the expected service life of the finer sized wires.

Types K (Chromel (nickel-10% chromium) (+) vs. Alumel (nickel-5% aluminum and silicon (-)) and T (Copper (+) vs. Constantan (nickel-45% copper) (-) thermocouples are often used below 0°C, but type J is not suitable for general low-temperature use because the positive thermo element (noted as JP) is composed of iron and thus is subject to rusting and embrittlement in moist atmospheres. Type K is more resistant to oxidation at elevated temperatures than types E, J and T and consequently it finds wide application at temperatures above 500°C. Type E has the highest thermopower above 0°C of any of the standardized types.

Type N, Nicrosil (nickel-14% chromium, silicon) (+) vs. Nisil (nickel-4% silicon, magnesium) (-). This type differs from type K by having silicon in both legs and containing magnesium in the negative leg. It was developed to be more stable (exhibit less calibration drift) than type K when used at temperatures above about 1000°C (1800°F). Both type N wires are similar in color and both are non-magnetic, so identification is usually made by gently heating the junction and observing the polarity of the resultant emf.

3.4 Noble metal types

Thermocouples employing platinum and platinum-rhodium alloys for their thermoelement (Noble-metal thermocouple types B, R and S) have been used for many years and exhibit a number of advantages over the base metal types. They are most resistant to oxidation, their thermoelements have higher melting points, and they have generally been found to be more reproducible at elevated temperatures in air. They are therefore used when higher accuracy and longer life is sought, though more experience with lower thermopowers.

Of all the standardized thermocouples, Type S, Platinum-10% rhodium (+) vs. Platinum (-) is the oldest and perhaps the most important. Type B, Platinum-30% rhodium (+) vs. Platinum-6% rhodium (-), was adopted as a standard type in the US in the late 1960s primarily to serve requirements in the 1200 to 1750°C range. At elevated temperatures, it offers superior mechanical strength and improved stability over types R and S, and it exhibits comparable thermopower. Its thermopower diminishes at lower temperatures and is vary small in the room-temperature range.

Identification of noble metal thermocouple wires is made difficult by the fact that all alloys are nearly identical in colour and all are non-magnetic. Sometimes it is possible to distinguish the positive wire from the negative one for types R or S by observing the 'limpness' of the wires. Pure platinum wires tend to be slightly more soft, or limp, while the rhodium-alloyed conductors are a little stiffer, enough so to permit identification. The differences, however, are subtle, and it is not possible to tell one rhodium alloy from another by these means. Proper connections for these thermocouples can be reliably determined by gently heating the junction and observing the resulting polarity on a sensitive indicator.

4. Solar harvest circuit design

The core of the harvesting module (solar panel) is the harvesting circuit, which draws power from the solar panels, manages energy storage, and routes power to the target system. The most important consideration in the design of this circuit is to maximize efficiency and there are several aspects to this. Solar panels have an optimal operating point that yields maximal power output. The harvesting circuit should ensure operation at (or near) this maximal power point, which is done by clamping the output terminals of the solar panel to a fixed voltage.

A DC-DC converter is used to provide a constant supply voltage to the embedded system. The choice of DC-DC converter depends on the operating voltage range of the particular battery used, as well as the supply voltage required by the target system. If the required supply voltage falls within the voltage range of the battery, a boost-buck converter is required, since the battery voltage will have to be increased or decreased depending on the state of the battery. However, if the supply voltage falls outside the battery's voltage range, either a boost converter or a buck converter is sufficient, which significantly improves power supply efficiency.

4.1 Material consideration for fabricating solar panel

4.1.1 Choice of composite material

A material for fabricating solar cells should be cheap to acquire and must be pure. Attempt on polymer and composite material based cell is a good development. Composites in general showed good physical properties and improved mechanical strength, 0-3 type super conducting composites with epoxy and phenolic thermosetting plastics have advantages of high toughness, superior abrasion, dimensional stability and heat, water and chemical resistance.

The composition of naturally occurring or pyrolytically obtained composite material (Bitumen), is complex but separation, by both physical and chemical methods, into different chemical groups has been made (Oyekunle, 1985). The fractions so obtained consist of asphaltic hydrocarbons (asphaltenes) viscous naphtheno-aromatic hydrocarbons (heavy oils), heterocyclic and polar compounds (resins). Asphaltenes are hard, non plastic high, molecular weight compounds ranging between 1200 and 200,000 and are thus responsible for temperature susceptibility (Gun, 1973).

4.1.2 Asphalt composition

135cl by volume of emulsified (Cutback) asphalt (mixture of bitumen and mineral aggregates of 0/5mm size) were sourced locally. The bitumen is heated in a container (hot-mix plant) and mixed thoroughly with aggregates to form asphalt concrete. The composition is as shown in table 1.

Material	Property
Mineral Aggregate Sieve Size	No. 40
Mixture type	A
Percentage Passing	0-8
Bituminous Material:	MC Liquid asphalt, MC 250

Table 1. Asphalt Properties and Compositions

4.1.3 Asphalt preparation

The mineral aggregates and bituminous material is in proportions by weight. The aggregate is ensured clean and surface dry before mixing. The mixing period is sufficient to produce a uniform mixture in which all aggregate particles are thoroughly coated. Asphalt cement content of mixes is an important physical characteristic and influences the performance life of asphaltic concrete. Too much asphalt cement results in mixture stability problems, while too little asphalt cement results in a mixture that is not durable, (Robert *et al.*, 1996; Gordon 1997).

4.1.4 Thermocouple material

Thermocouple: Type E thermocouple has good stability, highest sensitivity among the common metals and thus has high emf output. Based on ASTM set recommended upper temperature limits for various wire sizes, selected diameter for the E type thermocouple is AWG24, 0.51mm diameter. Upper temperature limits for E type is 427°C. Type E has the highest thermopower of 6.317mV/°C in the temperature range (0 -100) °C among any of the standardized types. The thermocouple properties are as shown in Table 2.

Property	Nickel- Chromium (Constantan)	Copper – Nickel (Chromel)
Composition	90% Ni, 10% Cr	60% Cu, 40% Ni
Thermal conductivity, k	22.7-w/m² °C	17.1-w/m² °C
Thermal diffusivity, α	61.2 x10⁴ m²/s	44.4x10⁴ m²/s
Density, ρ	8922 kg/ m²	8666kg/ m²
Electrical conductivity	58.14 x10⁶ m⁻¹ Ω⁻¹	-
Useful temperature range	-200 to 980	-
Total thermal conductivity	5.4 x10⁻⁴ W/ m² °C	
Diameter	0.51mm (0.00051m)	
Length	15mm (0.015m)	
Figure of merit	1.0 x 10⁻⁶	
Thermopower at (31-80)°C	3.116 mV/°C	

Table 2. Thermocouple Material Properties

4.1.5 Extension wire

Thermocouple alloy wire is recommended to be used always to connect a thermocouple sensor to the instrumentation to ensure accurate measurements. Due to the high cost of thermocouple wire, a copper wire was used with assurance of no significant change in the emf produced.

4.1.6 Heat source

The mixture is compressed into small pallets (0.5cm) with thermocouple junctions cascaded and embedded into it. A glass screen is to be provided to prevent escape of long wave radiation from the absorber surface.

4.2 Modeling the generator

The following assumptions are used in determining the open circuit voltage: The heat source is a flat plate collector, thus an assumption of maximum temperature of 80°C is considered for the heat source, a black body absorber. An ambient temperature of 31°C was considered for the isothermal block because ice baths are often inconvenient to maintain at 0 °C and not always practical.

4.2.1 Output voltage

To determine the output voltage X of the thermocouple at 31.0 °C and 80.0°C, the thermoelectric emf at 31.0°C is interpolated (S_{31} = 1.867 mV/°C) and the difference used to determine $S =$ S_{80} - S_{31} = 3.116 mV/°C. The output (open-circuit) voltage, V of the thermocouple junction is given by V = S (T_{80} –T_{31}) =   152.68 mV

4.2.2 Thermocouple (heat sink) reference junction

With accurate thermocouple measurements required, it is common practice to reference both legs to copper lead wire at the ice point so that copper leads may be connected to the emf readout instrument. This procedure avoids the generation of thermal emfs at the terminals of the readout instrument. The emf generated is dependent on a difference in temperature, so in order to make a measurement the reference must be known. The reference junction is placed in an ice water bath at a constant 0°C (32°F). Because ice baths are often inconvenient to maintain and not always practical, several alternate methods are often employed (Omega Engineering, 2001)

4.2.3 Solar cell configuration

Under bright sunlight, all silicon PV cells have an open circuit output of approximately 0.5V irrespective of cell surface area. The voltage is a function of the cell's physical composition, while amperage is affected by area of cell and the amount and intensity of light falling upon it. Increase in the voltage and amperage output of PV cells depends on the mode of connection of the cells in a module. For higher voltage, the cells are linked in series (net voltage is the sum of the individual voltages of the cells) (Figure 8a). The net current is however the same as that of a single cell. To boost amperage, the cells must be connected in parallel (Figure 8b)

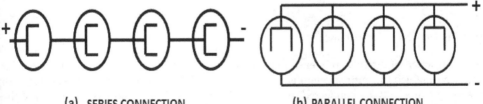

(a) SERIES CONNECTION   (b) PARALLEL CONNECTION

Fig. 8. Solar cell connections

4.3 Construction of thermocouple circuit

The thermocouple wires (Type E) made of different metal alloys (Nickel-Chromium copper-constantan) is joined together by soldering. The number of thermocouples required to generate an output voltage of 15V is required. Knowing the output voltage of one thermocouple (type E) given as 153mv (0.153v), dividing 15 by 0.153 to give 98.04 = 98 junctions. There are six modules with 15 junctions each (Fig 9). These thermocouples were joined together in series to form cascaded thermopile consisting of a number of thermocouples.

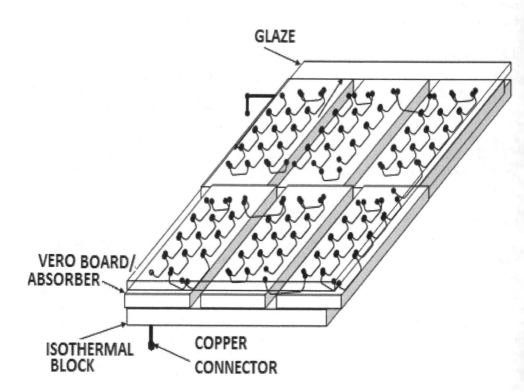

Fig. 9. Thermocouple solar panel

The construction of the solar power module was simple and convenient employing modular approach in which the entire system is divided into modules. The design is to generate high voltage, thus the cells are connected in series in the module. The voltage is a function of the cell's physical composition, while the amperage is affected not only by the area of the cell, but also by the amount and intensity of light falling upon it.

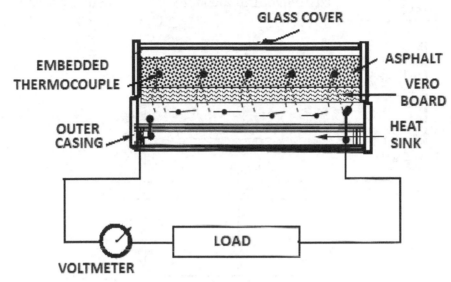

Fig. 10. A section through the solar panel

For high amperage, the cells must be connected in parallel. The net voltage is the sum of the individual voltages in the cell. Increase in the voltage and amperage output of the thermocouple cells depends upon the mode of connection in the module. The efficiency and power output requirement is determined by the power output of one thermocouple given by the equations above, the number of thermocouples in series, and the surface area of the solar cell were thus determined.

4.4 System fabrication

The entire system is divided into modules as shown in Figure 11. Vero board is used as the circuit boards for the solar panel and the charge control system.

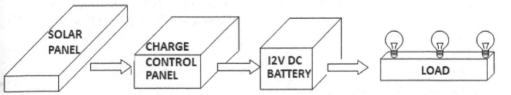

Fig. 11. Block diagram of the system

The charge control system uses the LED control charging system to charge a 12v lead Acid battery. An electrical diode, D_1 ensures unidirectional voltage flow when battery is under charge (Fig 10). For simplicity of construction and convenience the modular approach of constructing solar energy harvest modalities is used. Photovoltaic conversion provides the highest power density, which makes it the modality of choice to power an embedded system using reasonably small harvesting module.

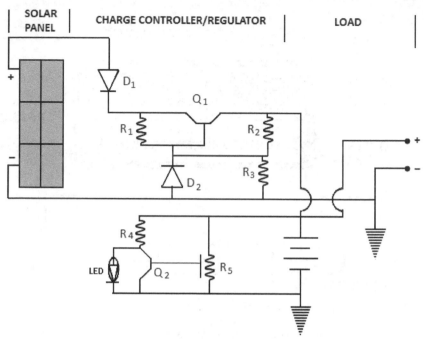

Fig. 11. Circuit Diagram of the solar power supply

The components of the electrical circuit and ratings are as follows:

D_1, D_2 = Diode (MA2J728 or MA3x704), Q_1, Q_2 = FET transistor (IRFZ44), R_1 = Resistor =220kΩ, R_2 = Resistor = 12kΩ, R_3 = Resistor = 2.7kΩ, R_4 = Resistor = 4.5kΩ, R_5 = Variable Resistor = 1000kΩ, LED (Green/Red), Battery = 12V Rechargeable Lead Acid.

Since the thermocouple array is expected to charge the battery on sunny days when output exceeds the load, but on cloudy days or at night, the load is expected to exceed the array output and drain the battery- Hence the array must be sized to ensure that the balance is positive and the battery is recharged when discharged. The array delivered an average daily output equal to the average daily system load (including all losses) plus approximately 10% to ensure that the battery is recharged.

5. System test result

5.1 Collector surface temperature

The daily total solar energy Q_t received per unit surface area of the absorber at the location (Ishiagu, South East Nigeria) as evaluated by Bello and Odey (2009) is 747.67 W/m^2. The useful components of the global solar radiation at the location are: direct solar radiation q_D = 680.67 W/m^2, diffuse solar radiation q_d = 64.21 W/m^2 and ground reflected radiation q_r = 2.34 W/m^2. The collector heat transfer coefficient between the absorber and cover expressed as the heat loss per unit area of the collector surface per temperature change is 3.06 W/m^2 °C. Total absorbed heat energy per unit surface area of absorber q_u = 592.43 W/m^2.

Measurements were taken on a clear day without cloud cover and surface temperatures were measured at five different spots every hour. According to measured temperature data, the average daily surface temperature increases with increase in sunshine hour reaching its peak between 1300hr and 1400hr (Fig 12) and then decline. On the average, 10hrs of sunshine hours is available per day, but for useful solar harvest, 8hrs of sunshine may is assumed because of difference in temperature between the collector surface and the ambient. The full sun (peak sun) hour value monitored at the site during raining season and dry season were found to be 4hrs and 5hrs respectively, agreeing with Onojo et al., (2004).

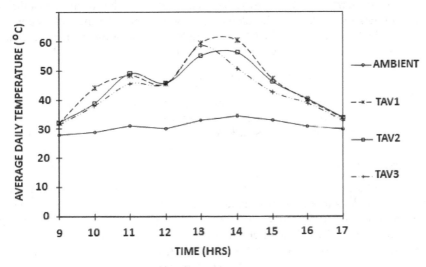

Fig. 12. Average daily temperature in collector

Three surface areas were used for the test as follows 0.6m², 1.0m² and 1.5m². There appeared to be no significant difference in spot temperatures in each of the surfaces per hour (Fig 12), it can be concluded that the surface temperature is independent of surface area. There appeared to be no significant difference in spot temperatures measured in each of the collectors per hour (T_{AV1}, T_{AV2}, T_{AV3}), hence, it can be concluded that the surface temperature is independent of surface area. The average cell surface area used in computation is 1.03 m² and the total surface area of the module is 6.2 m².

Material density constitutes to heat retention within the system and hence increases in surface temperature and higher potential difference. The collector compaction test shows that a densely packed material retains more heat and hence increases surface temperature, which obviously will produce higher potential difference.

5.1.1 Thermocouple sizing

The process of PV array sizing was utilized in determining the number of thermocouples to give the desired amount of electrical power required. To achieve this, the amount of electrical power required by the load and the amount of solar energy available at the site are necessary. The amount of ampere load energy demand required for a fixed load such as 2 DC T8 2ft fluorescent tube is 5.14Ah. Therefore the total demand in ampere-hours is 5.14Ah.

5.1.2 Battery size requirement and efficiency

When sizing the battery bank, the ampere-hours efficiency (columbic efficiency) of new battery is considered to be 100%. During the charge/discharge cycle of the battery, the battery is charged by receiving an input voltage from the thermocouple system and the same number of voltage is delivered at a lower output voltage. The battery efficiency (battery's voltaic efficiency) is expressed as the ratio of average voltage output to the average input voltage.

The daily load requirements determine the necessary battery bank capacity while the system voltage determines the battery bank voltage and the number of cells to be connected in series. The product of Daily Load (DL) requirement and the number of no-sun days (N) gives the total useable capacity (TUC) of the battery i.e.

$$Total\ Useable\ Capacity = DL \times N\ (Ah)$$

The ampere-hours efficiency (columbic efficiency) of new battery is considered to be 100%. The daily load requirement determines the necessary battery bank capacity. The total useable capacity (TUC) of the battery is 22.84Ah

5.1.3 Daily load requirement

An assumption of a lighting programme in poultry house where power is needed for four out of every seven days in a week was made, and the inverse relationship between voltage and amperage was used to determine the average daily current requirement by multiplying current by a factor of 4/7 to yield a net value in Ah, the average daily current required to satisfy the load demand of 5.14Ah as calculated from given relations. An average of 4½ hrs of full sunshine hours per day round the years is taken for a non-critical system. The thermocouple array load capable of generating the required load demand is obtained by dividing the average daily current requirement by 4.5.

$$Thermocouple\ load = average\ daily\ current\ requirement\ /4.5\ (A).$$

When a peak sunshine hour of 4.5 hours/day is required, the thermocouple array designed is capable of generating a measured 1.14A, capable of providing a glow continuously to satisfy the load demand of 5.14Ah. At increased number of sunshine hours above 4 ½ hours, more current generation is possible whereby the battery could be recharged.

5.1.4 The system conversion efficiency

The conversion efficiency is defined as the ratio of electrical power output and the heat flux through the entire TEG surface.

$$\varepsilon = \frac{P}{Q} = \frac{P}{\Delta T.A.h} \tag{38}$$

ΔT corresponds to the temperature difference between the hot and the cold side of the TEG, A is the TEG area and h is the overall heat transfer coefficient given by (the ratio of total thermal conductivity ($5.4 \times 10^{-4}\ W/m^2/^\circ C$) of the materials of the thermoelectric generators and the thickness (0.015m) of the TEG. The electrical power output (P=174.06 W h). The measured heat flux through the entire TEG surface is 10.94 W. The overall conversion efficiency of the system calculated is 15.91%. The cost of system production is estimated at average N20, 000.00

6. Summary

The conversion efficiency of the cell is 15%. This is comparable to other solar TEG system efficiency. The research work indicates the possibility of the utilization of asphalt bonded thermocouples to generate enough current for lighting programme in small scale agricultural undertaking such as poultry house illumination. The output voltage across the thermocouple generator can be increased to higher value enough to provide energy for other low thermal processes. Asphalt heat absorber will be a promising solar harvest cell when the surface is polished and made more sensitive to wider photon energy range (1.3-1.5eV) for increased efficiency.

From the economical point of view, there exists a huge discrepancy between the costs of commercial thermoelectric generators compare with asphalts embedded TEG. The commercial TEG is by a factor of 10 more expensive than the asphalt TEG. The properties of asphalt TEGs are comparable with that of commercial TEG, even though the asphalt TEG used in this study has a smaller area than most commercial TEGs, therefore more asphalt TEGs per unit area can be mounted for increased overall performance at a cheaper price. Further research on antireflection coatings and stacking of different cells with band gaps covering the incident energy of the photons would be a good attempt at achieving higher efficiency.

7. References

Afolabi M. O., Ajayi, R. I. and Siyanbola W. O., 2004. Photo-voltaic cells, efficiency and optimization. *Global J. pure & Applied Sc.* Vol 10. No.3 P 435-439.

Akarakiri J. B. and Ilori M. O., 2003. Application of photovoltaic technology in developing countries. *Nig. Jour. of Industrial & Systems Studies (NJISS)* vol.2 No.2

Angrist S. W., 1982. Direct Energy conversion. 4th Ed.

Bello S. R. and Odey S. O., 2009. *Development of Hot Water Solar Oven for Low Temperature Thermal Processes.* Leonardo Electronic Journal of Practices and Technologies ISSN 1583-1078 Issue 14, January-June 2009 p. 73-84

Champier D., Bedecarrats J. P., Rivaletto M., Strub F., 2009. Thermoelectric power generation from biomass cook stoves. Energy (2009) 1–8. doi:10.1016/ j.energy.2009.07.015 http//: www.elsevier.com/locate/energy

Coker J. O., 2004. Solar energy and its Applications in Nigeria. Short Communication, *Global Jour. pure & Applied Sc.* Vol. 10 No 1 P 223-225.

Dennis A. and Kulsum A., 1996. The Case for Solar Energy Investments. *World Bank Technical Paper Number 279.* Energy Series.

Duffie, J. A. and Beckman, W. A., 1976. Solar Energy Thermal Processes. John and Sons Pub NY.

Encyclopedia Britannica Inc., 1987: Thermocouples. Vol. 11 pp 250-251, Vol. 15 pp 226-227, Vol. 8 pp 193 William Benton publisher.

Gun R. B., 1973. Petroleum Bitumen (in Russian), khimiya, Moscow.

Goldsmid H. J., 1995. Conversion efficiency and figure-of-merit. In D. M. Rowe, editor, CRC Handbook of Thermoelectrics, pages 19-25. CRC Press,

Lertsatitthanakorn C., 2007. Electrical performance analysis and economic evaluation of combined biomass cook stove thermoelectric (BITE) generator. Bioresource Technology 2007; 98:1670–4.

Madueme T. C., 2002. Independent power producers and the power sector in Nigeria. *Nig.J. Ind & Sys. Studies (NJISS)* vol.1 no2. P.38-45.

Maycock D. Paul, Edward N., Stirewait, 1981. A guide to the Photovoltaic revolution. Rodale Press, Emmans, Pa.

Nolas G. S., Sharp J., and Goldsmid H. J. 2001. Thermoelectrics: Basic Principles and New Materials Development. Springer, Berlin,

Nuwayhid R. Y., Rowe D. M., Min G., 2003. Low cost stove-top thermoelectric generator for regions with unreliable electricity supply. Renewable Energy 2003; 28:205–22.

Nuwayhid R. Y., Shihadeh A, Ghaddar N., 2005. Development and testing of a domestic woodstove thermoelectric generator with natural convection cooling. Energy Conversion and Management 2005; 46:1631–43.

Okelola, O. E., Bamgbade., B. J. Balogun, R. B., Bello S. R., (2011). Qualitative Analysis of Rice Storage System in Yala Local Government Area of Cross River State. *Proc. of the 1st International Conference on Rice for food, Market and Development,* held in Abuja Nigeria. http://www.rice-africa.com/acceptedpapers/index.html#bv000030

Oladiran M. T., 1999. New and renewable Energy Education in Sub. Saharan Africa. *Proc. of renewable Energy Conf. Perth, Austria-energy 16.*

Omega Eng., 2001. Introduction to thermocouples. http://www.omega.com/techref/themointro.html /Omega Engineering. Date modified 8/12/2009.

Onojo O. Chukwudebe G. A. and Nwodo T. C., 2004. Development of solar power supply for domestic electricity. NJISS, Vol 3, No 2, 2004 pp 37-44

Oyekunle L. O., 1985. Effect of temperature on rheological properties of petroleum Bitumens. *Jour. Nig society chem. Engr.* Vol 4 pp 124-129.

Quinn T. J., 1983. Temperature, Academic press.

Richner P., Pedro D. G., Luís C. G. and David A., 2011. Experimental Results Analysis of the Energy Conversion Efficiency of Thermoelectric Generators. Electromechanical Engineering Department – Engineering Faculty University of Beira Interior Edifício 1 das Engenharias, Calçada do Lameiro, 6201-001 Covilhã (Portugal). http//www.312-richer.pdf . Date modified 8/12/2011.

Rowe D. M., 2009. Thermoelectric waste heat recovery as a renewable energy source. International Journal of Innovations in Energy Systems and Power 2006; 1(1).

Solar Electric Light fund (SELF), 2002. Solar Electricity and Renewable Energy Technology. *Washington D.C, solar Electric light Fund.*

Steven A. Moses, 2010. The Design and Construction of Two Experimental Setups to Measure Thermoelectric Properties of Novel Materials. A thesis submitted in partial fulfillment of the requirements for the degree of Bachelor of Science, Honors (Physics) at the University of Michigan

Suharta, H, Senam P. D. Satighm A. M. and Komarudin D., 1999. The social Acceptability of solar cooking in Indonesia. *Renewable energy* vol. 6 1151- 1154.

Zulovich J. M., 2005. Poultry Farm and Processing Plant Lighting. Published by University of Missouri extension

Permissions

The contributors of this book come from diverse backgrounds, making this book a truly international effort. This book will bring forth new frontiers with its revolutionizing research information and detailed analysis of the nascent developments around the world.

We would like to thank E. B. Babatunde, for lending his expertise to make the book truly unique. He has played a crucial role in the development of this book. Without his invaluable contribution this book wouldn't have been possible. He has made vital efforts to compile up to date information on the varied aspects of this subject to make this book a valuable addition to the collection of many professionals and students.

This book was conceptualized with the vision of imparting up-to-date information and advanced data in this field. To ensure the same, a matchless editorial board was set up. Every individual on the board went through rigorous rounds of assessment to prove their worth. After which they invested a large part of their time researching and compiling the most relevant data for our readers. Conferences and sessions were held from time to time between the editorial board and the contributing authors to present the data in the most comprehensible form. The editorial team has worked tirelessly to provide valuable and valid information to help people across the globe.

Every chapter published in this book has been scrutinized by our experts. Their significance has been extensively debated. The topics covered herein carry significant findings which will fuel the growth of the discipline. They may even be implemented as practical applications or may be referred to as a beginning point for another development. Chapters in this book were first published by InTech; hereby published with permission under the Creative Commons Attribution License or equivalent.

The editorial board has been involved in producing this book since its inception. They have spent rigorous hours researching and exploring the diverse topics which have resulted in the successful publishing of this book. They have passed on their knowledge of decades through this book. To expedite this challenging task, the publisher supported the team at every step. A small team of assistant editors was also appointed to further simplify the editing procedure and attain best results for the readers.

Our editorial team has been hand-picked from every corner of the world. Their multi-ethnicity adds dynamic inputs to the discussions which result in innovative outcomes. These outcomes are then further discussed with the researchers and contributors who give their valuable feedback and opinion regarding the same. The feedback is then collaborated with the researches and they are edited in a comprehensive manner to aid the understanding of the subject.

Apart from the editorial board, the designing team has also invested a significant amount of their time in understanding the subject and creating the most relevant covers. They scrutinized every image to scout for the most suitable representation of the subject and create an appropriate cover for the book.

The publishing team has been involved in this book since its early stages. They were actively engaged in every process, be it collecting the data, connecting with the contributors or procuring relevant information. The team has been an ardent support to the editorial, designing and production team. Their endless efforts to recruit the best for this project, has resulted in the accomplishment of this book. They are a veteran in the field of academics and their pool of knowledge is as vast as their experience in printing. Their expertise and guidance has proved useful at every step. Their uncompromising quality standards have made this book an exceptional effort. Their encouragement from time to time has been an inspiration for everyone.

The publisher and the editorial board hope that this book will prove to be a valuable piece of knowledge for researchers, students, practitioners and scholars across the globe.

List of Contributors

Jose Maria Cabeza Lainez
University of Seville, Spain

Oreste Boccia, Fabrizio Chella and Paolo Zazzini
D.S.S.A.R.R., University "G. D'Annunzio" Chieti, Pescara, Italy

Nilgün Sultan Yüceer
Çukurova University, Faculty of Engineering and Architecture, Balcali Adana, Turkey

Antonio Eduardo Hora Machado, Lidiaine Maria dos Santos, Karen Araújo Borges, Vinicius Alexandre Borges de Paiva, Paulo Souza Müller Jr., Danielle Fernanda de Melo Oliveira and Marcela Dias França
Universidade Federal de Uberlândia, Instituto de Química, Laboratório de Fotoquímica, Uberlândia, Minas Gerais, Brazil

Paulo dos Santos Batista
Universidade Federal de Goiás, Campus Catalão, Departamento de Química, Catalão, Goiás, Brazil

Manuel Gálvez, F. Javier Rodríguez and Emilio Bueno
Department of Electronics, Alcalá University, Spain

Qi Luo and Kartik B. Ariyur
Purdue University, USA

Flora Bougiatioti
Hellenic Open University, Greece

Benito Corona-Vasquez, Veronica Aurioles and Erick R. Bandala
Grupo de Investigación en Energía y Ambiente, Fundación Universidad de las Américas, Puebla, Santa Catarina Mártir, Cholula, México

Kanayama Kimio
Kitami Institute of Technology, Japan

Koga Shinya
Kyushu University, Japan

Baba Hiromu
Formerly, Kitami Institute of Technology, Japan

Sugawara Tomoyoshi
Marusho-giken Co., Ltd., Japan

R. S. Bello, K. A. Eke, R. B. Balogun and O. Okelola
Federal College of Agriculture, Ishiagu, Nigeria

S. O. Odey
Cross River University of Technology, Obubra, Cross River, Nigeria

M. A. Suleiman
Federal College of Agriculture, Jalingo, Nigeria

T. A. Adegbulugbe
Federal College of Agriculture, Moore Plantation, Ibadan, Nigeria

Printed in the USA
CPSIA information can be obtained
at www.ICGtesting.com
JSHW011418221024
72173JS00004B/581